FRUIT AND VEGETABLES

A CREATIVE STEP-BY-STEP GUIDE TO

FRUIT AND VEGETABLES

Author
Peter Blackburne-Maze

Photographer
Neil Sutherland

WHITECAP BOOKS

This edition published in 1997 by
Whitecap Books Ltd., 351 Lynn Avenue
North Vancouver, B.C., Canada V7J 2C4
CLB 4137
Printed in Singapore
ISBN 1-55110-285-4

Credits

Edited and designed: Ideas into Print
Photographs: Neil Sutherland
Typesetting: Ideas into Print and Ash Setting and Printing
Production Director: Gerald Hughes
Production: Ruth Arthur, Sally Connolly, Neil Randles,
Karen Staff, Jonathan Tickner

THE AUTHOR

Peter Blackburne-Maze grew fruit commercially for many
years and although he has widened his experience to take
in all aspects of horticulture, fruit remains his first love and
specialist subject. He spent 11 years in the agrochemical
business and is now a horticultural consultant and writer.
He is a regular contributor to a number of gardening
magazines and has several books on gardening to his
credit. These cover both vegetables and fruit. He is the
current chairman of the RHS Fruit Group and a member
of the RHS Fruit and Vegetable Committee.

THE PHOTOGRAPHER

Neil Sutherland has more than 25 years experience in a
wide range of photographic fields, including still-life,
portraiture, reportage, natural history, cookery, landscape
and travel. His work has been published in countless books
and magazines throughout the world.

Half-title page: 'King of the Pippins', a traditional apple
variety with small fruits that make it an ideal choice for
garden use. The creamy white flesh is firm and juicy.

Title page: Squashes and pumpkins are becoming
increasingly popular for food and decoration.

Copyright page: Enjoy the fruits of your labors: a bowl
of fresh, succulent, homegrown raspberries.

CONTENTS

Part One

GROWING VEGETABLES

Anyone who has grown their own vegetables will know the pleasure and satisfaction involved, especially when they are sitting on a plate in front of you. It is also extremely cost effective and, what is more, it takes you outdoors and gives you exercise. If you are a newcomer to the subject, there are a few things to consider. For example, which vegetables are you are going to grow and how large an area are you prepared to devote to them? Clearly this will be governed by the size of your garden and how much you like vegetables. As a rule, it is best to avoid crops such as maincrop potatoes and maybe sprouting broccoli that take up a lot of space and/or that remain in the ground for a long time. On the other hand, new potatoes are particularly valuable; one row takes up very little room and new potatoes straight from the garden taste quite unlike any other. Peas and beans are also good value where space is limited, as they crop over a long period. Other worthwhile choices are the winter and early spring vegetables, such as spring greens and spinach beet, which are ready when vegetables in the shops are at their most expensive. Any vegetables that are expensive to buy are always worth growing and so are any that you like but that are difficult to find in shops. Having made these preliminary decisions, the first job is to get the soil into a suitable condition for growing vegetables. This means having the right tools, including a line and dibble for planting out seedlings. Now you are ready to embark on an enjoyable and rewarding venture.

Left: *A flourishing vegetable garden.* **Right:** *An eggplant (aubergine) with its starry flowers.*

Digging the soil

Just as plowing is the farmer's way of moving the soil deeply in preparation for shallower cultivations and seed-sowing, digging is the basic cultivation of the gardener. You must dig any ground that is being cultivated for the first time or that has just been cleared of an old crop. A spade is the normal tool used for digging, but on heavy ground, a strong fork is much easier to use and just as effective. The choice depends on how well the soil holds together. If it falls through the prongs of a fork, then you should use a spade instead.

There are several reasons for digging. The main one is to break up the soil so that excess water can drain away and roots can penetrate in their search for water and nutrients. As the ground is broken up, air is able to reach down into it so that the roots can 'breathe'. Digging also creates a suitable tilth for sowing or planting. When you dig cultivated ground, you bury any existing weeds and weed seeds, which reduces the competition that faces new young plants in their all-important first weeks. Digging is also the most thorough and reliable way of burying bulky organic matter, such as garden compost, that is vital if the soil is to 'live' and support good crops.

1 When single digging a plot of land, use a line to mark out the position of the first trench. The trench should be 9-10in(23-25cm) wide.

2 Using a spade, scrape the surface rubbish and weeds off the marked area to give a clean surface. Work your way along in a methodical way.

3 Dig out the soil to a full spade's depth and barrow it away to the far end of the plot, where it will be used to fill in the final trench.

4 After digging out the trench, clean out the loose earth in the bottom. The trench is now ready to receive the soil from the second trench.

Using a fork

A fork has two main uses in preparing the soil. It is much easier to use than a spade when digging heavy clay soil and it does a splendid job when you want to break down land that has already been dug. For this, use the back of the fork to break up the clods into a finer tilth. (See page 14 for more details.)

5 *If there is not much debris on the surface, turn the soil from the second trench over and forward into the first trench without any scraping.*

6 *If there are any remains of the previous crop on the surface, be sure to turn each spadeful of soil over completely to bury the rubbish.*

Do not make any attempt to break up heavy ground. The rougher you leave it, the greater the surface area there will be for weathering.

7 *Keep working backwards down the plot, scraping if necessary and turning the soil over and forwards into the previous trench as before.*

8 *Always keep the trench open and clear so that there is enough room to accommodate the next 'row' of soil excavated from the previous trench.*

9 *Fill the final trench with the earth that was dug out from the first one. If digging a vegetable plot in the fall, leave it to weather until spring.*

Below: A fork is one of the most useful tools for breaking down previously dug soil into a tilth suitable for sowing or planting. Do not dig, but knock the clods apart with the back of the fork.

Basic cultivation

In order to grow worthwhile crops, you need at least a basic knowledge of soil cultivation and the tools required to carry it out. As well as a spade and fork (see pages 12-13) you will need a rake to break down the soil surface still further before sowing seeds. When sowing seeds or planting out seedlings, always use a garden line; never rely on drawing out a straight line by eye. Once the plants are growing, they will inevitably be challenged by weeds. The cheapest way of dealing with these is to hoe them out, using either a Dutch or a draw hoe. Never try to cultivate the ground when it is wet enough to stick to the tools and always clean your tools after use. All garden tools can be made of either ordinary or stainless steel. Stainless steel costs more, but does not go rusty. As a general rule, always buy the best tools you can afford; the better they are, the longer they will last.

If the soil conditions are not suitable for the plants you want to grow, your results will be poor. Within the soil the most important material is the organic matter. In nature this comes from decomposed vegetation, but in the garden you must add it in the form of garden compost or farmyard manure. Normal levels of organic matter contain only small amounts of the essential plant foods, so you will need to add these in the form of fertilizers. Under average garden conditions, add both bulky organic matter and plant foods to the soil at least once a year for it to remain in a suitable state for plants to grow well.

Above: After breaking down the clods with the back of a fork, you can make the tilth even finer in preparation for sowing by raking the surface of the soil backwards and forwards.

Food and water

Below: Apply a granular fertilizer to the ground before sowing or planting. Long-standing crops, such as runner beans, will need more feed once they start cropping.

Above: The Dutch hoe is very useful for removing weeds amongst the vegetables. Use it shallowly when the soil is dry and the weeds are small. Killing weeds is a top priority.

Above: The same principles apply to the draw hoe. Catch the weeds in good time, otherwise they will soon be competing with the crop for water, nutrients and space.

Below: *During very dry periods , use liquid feeds in place of granular ones. The plants use the food instantly and also get some water.*

Above: *Rake in fertilizers after applying them so that they dissolve and move down into the soil more quickly. You can use either organic or, as here, inorganic feeds.*

Above: *Most vegetables will require additional water in the summer. The best irrigation systems are trickle or perforated hoses, as the water is applied where it is needed.*

Left: *To grow vegetables on heavy soil, create semi-permanent beds 4ft(1.2m) wide. You can walk between the beds and thus avoid treading on the crops.*

Right: *Inter-cropping makes excellent use of growing space. The lettuces will be long gone by the time the sweetcorn needs the room to grow.*

Below: *All being well, your vegetable garden should look like this by midsummer. Not a weed in sight and everything growing away.*

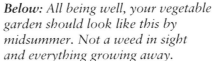

Making garden compost

Garden compost is one of the most common and effective forms of bulky organic matter that you can add to the ground, and it is also the cheapest. It is made from plant remains that are continually becoming available in every garden, including weeds, spent vegetables, lawn mowings, hedge clippings, flower stalks, leaves and soft prunings. You can also add sawdust, straw and hay, pet litter and bulky animal manures. Household waste, such as potato peelings, cabbage leaves, old cut flowers and tea leaves and a host of other things can all be added to make good garden compost. All these items are of a vegetable origin; never add anything of a meaty nature. The secret of making good garden compost is to use a good mixture of raw materials: plenty of soft vegetation together with a high proportion of woodier things. Autumn flower stalks, woody prunings, cabbage and other brassica stalks are all first-rate, but be sure to chop them up first with a spade or, better still, shred them mechanically. The choice between a compost heap or composting bin rests largely on the amount of raw material available. If there is plenty, a heap is better. However, where raw material is limited, bins are better. Small heaps never make good compost. Ideally, fill up bins in one operation; they quickly heat up and make the best compost.

1 When starting a new heap or bin, try to arrange it so that the first batch of material is coarse and woody, such as these rose prunings. It ensures that the heap will have good drainage and aeration.

2 Keep adding more coarse material until you build up a foundation layer of vegetation about 6in(15cm) deep in the bottom of the bin or heap, once you have firmed it down.

3 Now you can start to add soft materials as well, such as these grass mowings, which will heat up splendidly.

Shredding thorny prunings renders them harmless.

Using a shredder

Some of the best raw materials are woody and cannot go directly onto the compost heap. With a shredder you can make sure that none of this valuable vegetation is wasted. Where it is allowed, you could burn prunings instead and put the ashes on the heap, but this is second best.

Thorny prunings after they have been put through an electric shredder.

Proprietary activators contain nitrogen together with an agent that reduces acidity. The microorganisms prefer an alkaline environment.

4 With the shredded prunings, the heap is about 10in(25cm) deep. This is the time to sprinkle on an activator. It will add nitrogen to the raw materials, helping the microorganisms to break them down into rich garden compost.

5 Continue adding more raw material. These potato tops are soft and full of water so, ideally, you should follow them with rougher material to keep the heap open.

6 Another helping of shredded prunings would be suitable and will also assist the decomposition process by aerating the heap. This, in turn, leads to heating up.

9 The result is humus-rich, fibrous garden compost that will improve your soil physically and chemically, either as a mulch or when dug in.

7 With this type of bin you add slats as the heap rises. The gaps are for ventilation. Do not forget to add more activator for every 10in(25cm) or so increase in depth of vegetation.

8 An important part of good composting is to stop the generated heat escaping. Any substantial covering will do this and it will also stop rain from cooling the heap.

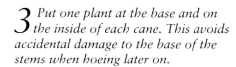

Climbing beans

1 *Fill 3.5in (9cm) pots with seed or multi-purpose potting mix. Make a small hole and put one seed in each pot, pointed end down, with the top of the seed 1in (2.5cm) below the surface.*

Climbing French and runner beans are two of the most valuable summer vegetables in the garden. They can start cropping in midsummer and, with reasonable upkeep, will carry on until the first fall frosts. Treat both as half-hardy annuals and do not sow them outdoors until mid-spring, by which time the soil will have warmed up and there is little risk of frost. Alternatively, sow them in a greenhouse or frame up to a month earlier, before planting them outside in early summer when all risk of frost has gone. Sow the seeds singly in 3.5in (9cm) pots of sifted soil, seed or potting mixture. Because both vegetables are climbers, provide a support system of strings or sticks, tall enough to allow the beans to climb to about 6ft(2m). The most convenient system is a double row of 8ft(2.5m) canes pushed into the ground. Tie the tops of opposite canes together across the row. Sometimes, runner beans have difficulty in setting the beans. You can improve their chances by spraying the flowers with water most evenings during flowering. Climbing French beans do not have this problem.

2 *Push four 8ft(2.5m) canes into the ground about 1ft(30cm) apart and tie into a wigwam. Remove each plant from its pot, holding it gently by the stem.*

3 *Put one plant at the base and on the inside of each cane. This avoids accidental damage to the base of the stems when hoeing later on.*

4 *Water the plants thoroughly. If the soil is dry, each plant will need at least one gallon(4 liters) of water to help it grow away quickly.*

5 When the plants are about 12in (30cm) tall, tie the main stem to the cane with soft string. After this, the plants will be able to climb up the canes unaided.

6 If the garden does not lend itself to growing vegetables in rows, a block of four wigwams is a useful alternative strategy. These plants have started to flower and look healthy.

7 When plants reach the top of the canes, pinch out the tops. This encourages side shoots to form and keeps the beans at a pickable height.

8 An embryo runner bean amongst the flowers. Encourage good pollination and setting by regular overhead spraying in the evening.

Below: Except for their climbing habit and the varieties, climbing French beans are the same as dwarf French beans. They are not as early-maturing as the dwarf varieties, but carry a heavier crop for a given area.

9 A good crop of runner beans is one of the most valuable vegetables in the garden. They can be eaten at once or frozen for winter use.

Dwarf beans

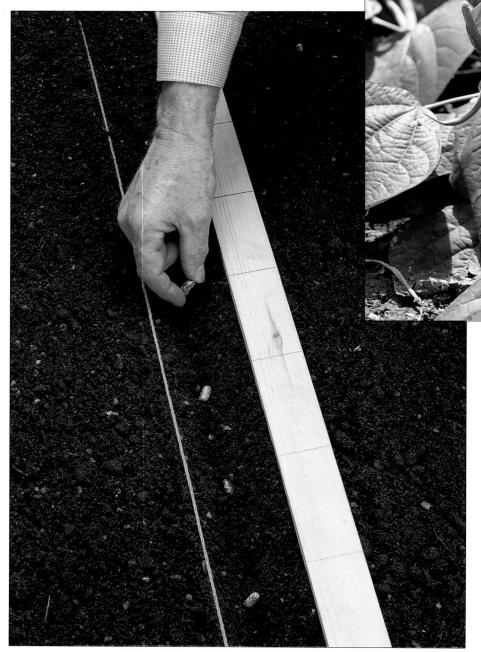

1 *Being comparatively large, French bean seeds are easy to sow. You can even soak them in water overnight before sowing to soften the coats.*

2 *Sow French beans outdoors in shallow drills or dibble holes 1.5-2in (4-5cm) deep, with the seeds planted 2-4in (5-10cm) apart.*

Dwarf French beans are one of the first spring-sown vegetables to mature in the summer. They are usually sown before runner beans so that they mature ahead of them. There are two varieties: the traditional ones with fairly flat pods and the modern 'pencil-podded' kind that are stringless and, therefore, good for freezing. Do not sow these fatter varieties until late spring or early summer, when temperatures are higher. Then sow them in succession. The older, flatter varieties, on the other hand, are hardier and more reliable in poor summers. Although dwarf French beans may be sown earlier if cloches or a plastic tunnel are available, they are seldom raised under glass for transplanting outside later on. Although dwarf French beans are not as liable as the climbing sorts to wind damage in exposed gardens, they still do best in a warm and sunny position. The most important thing to remember about all kinds of beans whose whole pods are eaten (as opposed to broad beans, which are shelled) is that they should be picked when still young. If they are left until they reach their maximum size, they will be tough and uneatable. All varieties will freeze well but some are better than others when cooked after thawing. If your main aim is to freeze them for later use, choose varieties that are specifically recommended for freezing. These will normally be the fatter-podded ones.

3 To maintain this standard of cropping throughout the season, be sure to give the plants plenty of water, especially in a dry year. Dwarf French and dwarf runner beans tend to be quite firmly joined to the plant. When they are ready to harvest, cut them off if necessary, but do not pull them.

Dwarf runner beans

Dwarf runner beans are often grown in small gardens where space is at a premium. They seldom crop as heavily as the climbers, but they still mature over a long period. In fact, although there are dwarf varieties of runner bean, you can keep climbing varieties small and bushy by pinching back all vigorous shoots as soon as you see them. Like their climbing relations, dwarf varieties may be sown under glass for planting outside when the risk of frost has passed. This variety is called 'Pickwick'. The plants are the same size as dwarf French beans.

Right: *The traditional type of dwarf French bean, such as 'The Prince', has flatter and wider pods than modern pencil-shaped varieties. It is also hardier, which allows earlier sowing.*

Broad beans in the ground

Broad beans are one of the hardiest and most easily grown vegetables and the first that many children grow. There are two main types of broad bean: those that you sow in late fall (Aquadulce type) and those that are best left until the late winter or early spring (Windsor and Longpod types). There are now dwarf varieties, which are particularly good for growing in small gardens or even in containers, such as growing bags. The advantages of fall sowing are that the plants crop a little earlier in the summer and are less likely to be attacked by blackfly (the black bean aphid). Another way to discourage blackfly is to nip out the tops of the plants when the bottom flower truss has set beans. You can cook the tops like spinach and these make a pleasant, if not exciting, change. Broad beans (except the dwarf varieties) grow into tall plants; most will reach 3-5ft(90-150cm) and therefore need support. The simplest method is to drive in a stout pole at the corners of each block of plants and stretch twine around them. This prevents the plants being blown about and also makes them easier to pick. The first beans will mature in early summer; never be tempted to wait until the beans within the pods are becoming large. Always pick them when they are still small and tender.

1 Sow the seeds singly in holes or in drills 1.5-2in(4-5cm) deep. Allow 18in(45cm) between the rows. Use a garden line to make the row straight.

2 Plant the seeds about 9in(23cm) apart. Use a trowel, which you have already measured, to determine where the next seed should go in.

3 Beans sown in early winter benefit from weather protection in early spring when they are growing again. Cloches also advance the crop.

4 All broad beans need support once they are 12in(30cm) tall. Run garden twine, looped around canes, around the whole group of plants.

5 *Once the lowest flowers on the stem have formed little beans, nip out the plant tips. This directs the plant's energy into bean formation.*

6 *Only remove the tip of the stem, leaving developing flowers intact to produce beans. Nipping out also discourages the attentions of blackfly.*

Above: *The reward of attentive gardening: a fine crop of beans for immediate use or for freezing. Do not let them become tough and black-eyed.*

7 *The young pods are growing well and filling nicely. Do not allow the plants to run short of water and treat them against blackfly if necessary.*

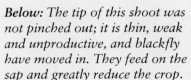

Below: *The tip of this shoot was not pinched out; it is thin, weak and unproductive, and blackfly have moved in. They feed on the sap and greatly reduce the crop.*

Broad beans in a tub

Broad beans are so easy to cultivate that there really is no reason at all why you should not grow them, even if you have insufficient space in the garden, as they make splendid plants for growing in containers of all sorts on a sunny patio. The main thing to remember is that you must stick to the dwarf-growing varieties. These are short and sturdy and generally need no support, whereas the normal tall ones do. Choose 'The Sutton' for this type of gardening. The other important point is that you must never grow broad beans under cover once they are more than about 4in(10cm) tall. Any extra warmth will draw them up, so that very soon they become leggy and topple over. Make sure you keep them outdoors, so that the flowers are adequately pollinated by bees.

Because the dwarf varieties are not sufficiently hardy to withstand the winter outdoors, and will become drawn if grown in any heat, you should sow them in the early spring. Unfortunately, this means that they are as susceptible to attacks of blackfly in the summer as are all the other spring varieties (Longpod and Windsor types). Luckily, this pest is easy to control. Besides removing the tip of each shoot once young beans are forming, spray any affected plants with a suitable chemical that will destroy aphids (greenfly and blackfly). Any systemic insecticide will do this, but a spray containing just pirimicarb is more acceptable as it only kills the aphids and nothing else.

1 Fill the container almost full with a good multipurpose potting mixture. Firm it down gently at this stage to prevent later compaction.

2 This is a dwarf variety 'The Sutton'. Sow the seeds individually on the surface about 4in(10cm) apart. Taller varieties are not suitable for containers.

3 Push the seeds into the potting mixture with your finger so that they end up 1.5-2in (4-5cm) deep. Move some soil back over the seeds and firm in gently.

Above: Children enjoy planting the large seeds of broad beans. The seeds are easy to handle and you can expect them all to germinate.

4 *Water thoroughly. Do the initial watering in easy stages but do not stop until some drains out of the base.*

5 *Now you can see the benefits of growing only dwarf varieties. The plants stay short and are much better suited to growing in containers.*

6 *The young plants growing away well and in flower. When grown outside (cooler than under cover), this variety should need no support.*

Garden peas

Garden peas have always been a favorite crop with gardeners. They are relatively easy to grow, are available throughout the summer and are splendid for freezing. Although the true garden pea is the most widely grown, you can also obtain seed for several other types. The petit pois is small-seeded and extremely sweet; with mange tout, or edible-podded, varieties (snap and sugar snap peas), the pods are eaten whole and not shelled. The 'asparagus', or 'winged', pea is also eaten whole, but is very unpealike with its faint asparagus flavor. Peas vary greatly in height and all of them, even the shorter varieties, are easier to manage if you provide them with some support to climb up. This can take the form of twiggy sticks or plastic or wire netting.

The unusual types of pea, such as asparagus pea and mange tout, are sown exclusively in the spring. Gardeners used to make an early sowing of true garden peas in the fall in much the same way as we do with broad beans today. However, there is little advantage in this over the more usual spring sowing. As a rule, the last sowing of garden peas should be no later than early summer, otherwise they are unlikely to mature before the fall frosts arrive. Strangely enough, the later sowings should be of early varieties because these are the quickest to mature.

Many gardeners grow their peas using the trench method. This involves digging out a trench in the winter and filling it with well-rotted garden compost or manure. Put the soil back and sow the peas in this soil later on. This method can result in excellent crops, especially where the natural soil is of a poor quality.

Left: *A well-filled pod of fresh peas from your garden - a tasty reward for your efforts. Be sure to keep the plants well watered, though, or they will stop cropping.*

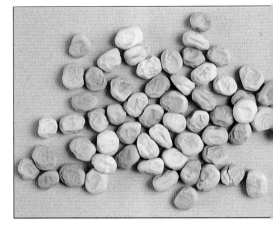

1 Pea seeds are nice and large and easy to handle. They look just like dried peas which, in fact, is what they are. Do not sow them too early as they may rot in cold, wet ground.

2 There are several ways of sowing and growing peas, but the simplest is in trenches about 8-10in (20-25cm) wide and 1-2in(2.5-5cm) deep.

3 Scatter the seeds about 2in(5cm) apart. This very arbitrary spacing allows a good, thick stand of plants to develop without being overcrowded.

Below: *Cook and eat the whole pod of sugar snap peas when the peas are almost full sized. They can also be used podded, like garden peas.*

Above: *Mange tout varieties are picked when the pods are still small, flat and tender, with the peas only just visible.*

4 Cover the seeds so that they are no more than 2in(5cm) deep. Even coverage is important so that all the seeds germinate more or less together.

5 Firm the soil down with the flat of a rake. This ensures that each seed is in contact with the soil and that the soil stays moist to help germination.

6 Most pea varieties need support. Sticks are the traditional choice, but wire netting is handier, longer lasting and easier to obtain.

Onions and shallots

You can grow onions either from seed or from 'sets'. Sets are the least troublesome way of growing onions. They are the size of baby pickling onions and are planted in the garden normally in mid-spring, although some are suitable for fall planting. Once planted, these miniature onions will grow to the full size. You can sow onion seeds either in late winter in a heated greenhouse for planting out later on or, if you choose the right varieties, you may sow the seeds outdoors in the early fall or in mid-spring. They like a firm and fine seedbed. If you store them well, onions grown one year will keep until the spring of the following year. Dig them up when the tops have died down in the early fall, clean them, dry them off thoroughly and store them in a cold shed. If you keep them indoors, they will start growing and will be useless.

Shallots are very similar to onions but with a milder flavor. They are normally grown from offsets planted outside in late winter or early spring. Once the bulbs are planted, prevent the birds from pulling them out before they have formed roots by threading black cotton along the rows. Each shallot planted will split up into five or six new bulbs. Harvest these in summer when the tops have died down. Onions, shallots and the closely related leeks, garlic and chives, may fall victim to a very unpleasant, damaging soil-borne fungus disease, white rot of onions. This attacks the developing plant at soil level, where it causes the white, feathery, stinking rot to appear. If you spot yellowing plants that have stopped growing, dig them up immediately and burn or throw them away, but never compost them. There is no reliable chemical control for white rot, but do not grow susceptible vegetables in that position for several years.

Above: To grow onions from sets plant them so that just the extreme tip is showing. If there is much more on view, birds will pull them up.

Below: A crop of good-sized 'Kelsae' onions in the making. They thrive on this firm, sandy ground and may even achieve a record-breaking size.

Right: Shallots (at left) and onions like the same growing conditions, so make good companions. Here, both are flourishing in early summer. Shallots were traditionally planted on the shortest day and harvested on the longest one, but this is not critical.

Below: *Onions and related crops, such as leeks, shallots, garlic and chives, have similar-looking seeds and seedlings. Label the rows straight after sowing to avoid any confusion.*

Above: *When growing onions from seed, make sure that they have a firm, fine seedbed. They need this to produce the best-quality bulbs for keeping.*

Right: *Salad, or green, onions are a favorite in summer. They are easy to grow from both spring and fall sowings. Use them before there is much of a swelling at the base. White Lisbon is a first-rate variety.*

Below: *Onions are also highly ornamental and come in a range of colors and sizes. It must be said, though, that the normal golden color has the greatest number of varieties and the highest quality for keeping. The highly colored ones are more of a novelty.*

Among shallots, 'Dutch Yellow' is excellent and 'Hative de Niort' has large, well-shaped bulbs.

Red varieties of onions include 'Red Baron' and 'Red Brunswick'.

There are many golden-skinned onions, including 'Hygro', 'Rijnsburger' and 'Sturon'.

Raising and growing leeks

Although they are not to everyone's taste, leeks do provide a welcome change from the incessant winter cabbages. Like onions, you can sow them either in a heated greenhouse in mid- to late winter or outside in the early to mid-spring. In either event, plant the seedlings outside in their final position when they are about as thick as pencils. To grow those tall, white-stemmed specimens that are the envy of everyone at produce shows, you have to sow early and plant them in the base of a specially dug trench. Later, you wrap newspaper around the young leeks and gradually fill up the trench so that the stems are kept white. To grow normal leeks, however, plant out the seedlings in early to midsummer, but not in trenches. Make holes 6in(15cm) deep and 6in(15cm) apart and, having trimmed the roots back to about 0.5in(1.25cm) long, drop a plant into each hole. No firming in is needed, you merely pour water into the hole. This covers the roots with soil and ensures that the little plants will flourish. Depending on the varieties and when you plant them, you can have leeks ready for digging up from late fall to early spring. The choice ranges from the older 'King Richard', one of the earliest, to the midseason 'Swiss Giant - Albinstar' and the late 'Cortina'. The real leek fanatic can be digging up 'Blauwgroene Winter - Alaska' well into spring.

Leek transplants before and after trimming.

5 *Trimming not only encourages strong new roots, but the smaller leaf area also makes drying out much less likely.*

1 *Sow leek seeds in early spring, either in nursery rows in the garden or in a small pot of seed-sowing mixture in the greenhouse.*

2 *Transplant leeks sown outdoors when they are about as thick as a pencil. Smaller and larger ones do not establish and grow on as well.*

3 *Trim the roots off the transplants, leaving less than 0.5in(1.25cm). Young leeks look very like spring onions and may be used as such.*

4 *Similarly, trim back the tops to leave each transplant about 6-8in (15-20cm) long. This reduces water loss and speeds up establishment.*

6 Make a hole for the roots and drop each transplant in so that a third to half is buried. Do not push the soil back into the hole.

7 Pour water into each hole until it is full up. This ensures that the soil is fully moist and some of it is washed down around the roots.

8 A batch of even and well-grown leeks a few days after transplanting. One or two outside leaves die off, but this is quite normal.

Growing exhibition leeks

1 To grow leeks of exhibition quality, sow two or three seeds in special paper pots in late winter.

2 Push the seeds down 0.5in (1.25cm) and fill in the hole above them with the mixture.

3 Water the pots well and stand them in a warm place in the greenhouse or conservatory or on a sunny windowsill for the seeds to germinate. This will normally take one to two weeks, according to the temperature.

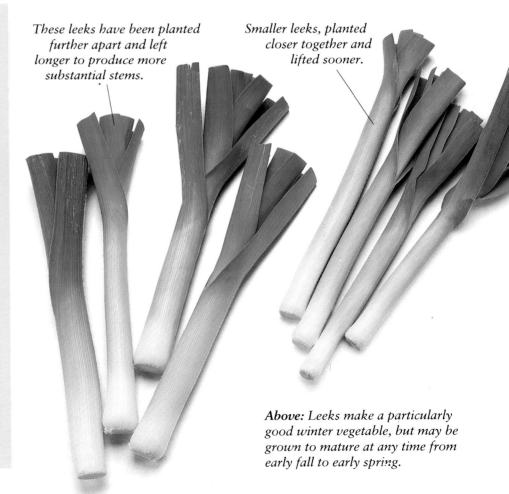

These leeks have been planted further apart and left longer to produce more substantial stems.

Smaller leeks, planted closer together and lifted sooner.

Above: Leeks make a particularly good winter vegetable, but may be grown to mature at any time from early fall to early spring.

Sowing lettuces

Lettuces can be raised in two ways, either by sowing them directly into the ground in the place where they are to mature (in situ) or by raising them under protection for planting out when it is warmer. The latter method is mainly used for the hearting varieties - iceberg, butterhead and cos - whereas the leaf varieties, which include many of the exotics, are usually sown in situ. All-year-round lettuces can be a complicated crop, but are quite easy to grow if you have the right conditions and varieties. When sowing lettuces in a greenhouse for planting out later, be sure to stop planting out in late spring and only sow lettuces in situ. If you do continue to plant them out, lettuces usually bolt (produce a flowerhead) because the check to their growth initiates flowering. Furthermore, seed germination is unreliable, or even stops, at temperatures above the mid 70s°F (mid 20s°C). It often pays to sow summer-maturing lettuces in semi-shade, where the temperature is likely to be lower.

1 Fill a suitable size pot with a good-quality seed sowing or multipurpose potting mix. Gently level and firm the surface of the soil.

Above: *Lettuce seeds are either white or almost black, depending on the variety. If you buy a mixture of varieties, you will often find both colors; do not let this worry you.*

Enjoying lettuces all year round

Late spring to early summer: Sow in heat mid- to late winter, plant outside mid- to late spring.
Early to midsummer: Sow the seeds in succession outdoors in spring.
Midsummer to mid-fall: Sow in succession outdoors late spring to late summer.
Early fall to early winter: Sow outdoors in late summer.
Fall and winter: Sow in cold frame in early and mid-fall, transplant seedlings to frost-free greenhouse.
Mid- and late spring: Sow the seed outdoors in late summer and early fall.
Use recommended varieties.

2 To aid seed germination, dampen the soil with a fine-rose before sowing until water runs out of the holes in the base.

3 Sprinkle the seeds evenly and thinly over the surface of the soil. If you sow too thickly, the seedlings will be lanky and weak.

4 Cover the seeds with sifted potting mix. Firm the surface lightly to ensure that the seeds are in contact with the soil.

5 *If you prefer to water after sowing, stand the pot in a bowl of water until moisture soaks up to the surface.*

6 *Cover the pot with a sheet of glass or plastic to maintain a damp atmosphere. Label it and stand it in a warm, but not too sunny, place.*

Sowing outdoors

You can sow lettuce seeds directly into well-prepared ground. These rows of seedlings are sufficiently well advanced to be thinned and singled to, say, 10in(25cm) apart. Fill any gaps in the row with the thinnings.

7 *At this stage, the seedlings are ready for pricking out. Take out a clump at a time and plant singly into a seed tray filled with potting mixture.*

8 *Make a dibble hole large enough to take the root system comfortably, put the roots in the hole and firm in the sides so that the seedling is self-supporting.*

Pricking out seedlings into individual pots

If you need only a few lettuce plants at a time, prick out the seedlings into individual peat or composition pots. Leave the seedlings in these pots when planting them out to avoid disturbing the root system. Always water thoroughly before planting out.

9 *There is no firm rule for spacing out the seedlings; 1.5-2in(4-5cm) is about right.*

10 *Water the seedlings thoroughly with a fine-rosed watering can. Plant them out after about one month and then start feeding them.*

Lettuce varieties

Although we used to consider lettuces as a summer crop, we can now grow and buy them throughout the year. Nor are they just of the traditional kind. The crispheads overtook the softer butterheads in popularity long ago, but we also have the cos varieties, together with the newer 'cut-and-come-again' leaf varieties and the colorful 'exotics'. Not surprisingly, lettuces vary in the planting conditions they require to flourish (see pages 32-33). According to the type and variety, they also vary in their spacing requirements. Space butterheads at 9-10in(23-25cm) apart, with 11-12in(28-30cm) between the rows. Icebergs prefer more room, so allow 15in(38cm) between each plant and between the rows. Cos and leaf lettuces can be planted slightly closer at 14in(35cm) square. It is always best to go by the spacing recommended on the seed packet, but follow these general guidelines if you are in any doubt. A good tip is to plant or thin at half the recommended spacing. Later, you can use alternate plants for an early crop.

Below: The cos is one of the heaviest lettuces. It also takes up a great deal of room, but is first-rate for a large family. 'Lobjoits' is an excellent variety.

Above: The 'exotics' are the latest things in lettuces. They are usually of the 'cut-and-come-again' sort, such as this frilly-leaved 'Lollo Bionda'.

Lettuce varieties

Summer-maturing (outdoors):
Butterhead: 'Avondefiance'
Crisphead: 'Saladin'

Summer leaf lettuce:
'Salad Bowl' and 'Red Salad Bowl', 'Lollo Rossa'

Winter leaf lettuce:
'Winter Density'

In the greenhouse:
Butterhead: 'Novita'
Crisphead: 'Kellys'

Above: Lettuces can be grown all the year round with protection. 'Kweik' is fine in an unheated greenhouse.

Exotic lettuces

The exotics are pretty enough to be grown in flower borders to very good effect. They carry on growing for many weeks; just pull off the leaves as you want them. Hearting varieties, such as the dark red-tinged 'Pablo', are better grown in the vegetable garden, as with these you cut the whole plant as required. They can all be sown where they are to mature and then singled. The 'leaf' varieties tend to have a stronger taste than the hearting ones.

Above: This collection of exotics includes 'Lollo Rossa', 'Lollo Bionda' and 'Red Salad Bowl'. All are easily grown in gardens.

Above: The popular 'Red Salad Bowl' goes well with 'Green Salad Bowl'. Both are 'leaf' lettuces that do not form a heart.

Left: This crisp 'Iceberg' type of lettuce is much heavier and denser than the butterhead sort, with its much softer leaves. There are a great many varieties of each to choose from.

Spring and summer cabbages

1 *Cabbage, cauliflower, Brussels sprouts, etc., have very similar seeds and seedlings. Label the rows straight after sowing or planting out.*

2 *Make a drill, or groove, in the ground about 0.5in (1.25cm) deep using a stick. Be sure to use a line for straightness. Failing this, use a draw hoe but do not go any deeper than is necessary.*

3 *Sow the seeds evenly and thinly so that they produce strong, stocky seedlings that will develop into good plants. If you sow the seed too thickly, the resulting seedlings will be crowded together and drawn.*

Fall and winter cabbage varieties take up rather a lot of space and for a considerable length of time, so it is more practical to consider the spring and summer varieties here. Although you would normally think of spring cabbages as being the first to mature in the year, in fact, a sowing of the summer cabbage 'Hispi' in a heated greenhouse in late winter will mature earlier. Harden off the seedlings and plant them outside in mid-spring 18in(45cm) apart when they are large enough. Alternatively, or if you do not have a heated greenhouse, sow the seed outside from early to mid-spring, plant out the seedlings in late spring and eat the result in midsummer. This, of course, would be later than spring cabbages. Successional sowings and later varieties will carry on the cropping season. Spring cabbage and spring greens are the most satisfying of vegetables. The 'greens' are simply immature spring cabbages and they are some of the first vegetables to be ready once the winter is over. The primary crop is spring cabbages, which are sown in mid- to late summer. However, if you transplant the resulting seedlings in early to mid-fall at half the spacing recommended for cabbages, you can cut alternate plants as greens when they are large enough and are touching in the row in spring. Leave the remaining plants to mature into fully fledged spring cabbages. Summer cabbages can be something of a mixed blessing, but just a few can make a welcome change in summer from the normal run of peas and beans. Beware, though, because come the fall, many vegetable gardens are 'wall-to-wall' cabbages and savoys until the winter is over. 'Durham Early' and 'Spring Hero' are good spring cabbages and 'Hispi' and 'Minicole' are recommended as summer varieties. The caterpillars of the large and small white butterflies can be devastating pests. At the first sign of the white butterflies in summer, you should spray the plants with a suitable insecticide, such as permethrin, or a biological one based on *Bacillus thurigiensis,* to avoid problems later on.

4 *A row of evenly germinating seedlings will follow if all the seeds are covered with soil to the same depth. The resulting seedlings will all be ready for planting out together. Uneven sowing leads to uneven seedlings.*

5 *A good way to mark where the plants are to go is to stand a pot at each station. Use a garden line and measuring stick to establish the exact positions for each pot.*

This is spring cabbage. The shape has no connection with the time of maturity, but pointed cabbages make the best spring greens.

Anyone would be proud of growing summer cabbages such as this one.

6 *You can use a dibble or, as here, a trowel for planting out the young cabbages. Make the hole deep enough for the root system to be vertical or the roots will not grow downwards.*

7 *Always plant firmly. This ensures that the roots will quickly start growing and the plant soon becomes established without first drying out.*

8 *A thorough watering is essential, even when the soil is moist. It means that the plant will receive only a slight check. Keep watering if necessary.*

9 *A beautifully even row of a first-rate summer cabbage, namely 'Hispi'. This variety can also be grown as a spring cabbage by sowing it in a heated greenhouse in late winter.*

Squashes and pumpkins

The cucurbit family, to which squashes and pumpkins belong, is a large one that includes marrows, courgettes (zucchini), cucumbers, gourds and melons. Their cultivation dates back about 7000 years. While marrows and cucumbers are common enough in Europe, squashes and pumpkins are far more widely grown in America, though they are spreading fast. As far as gardeners are concerned, there are two distinct types of squashes: summer and winter. Marrows and courgettes are a form of summer squash; pumpkins of winter squash. Although their cultivation is similar, summer squashes are essentially used when they are ready, but winter squashes are sown later and stored, when ripe, for use in winter. Harvest summer squashes well before they are old and tough, but leave winter squashes to mature. As a rough guide, winter squashes need about four months from sowing to maturity; summer squashes, somewhat less. Both may be of either bush or trailing habit. Because of the longer summer required for winter squashes, they are more popular in America than Europe, which favors the summer squash. Raise the plants in the same way as courgettes (page 40) from mid- to late spring sowings under cover and plant them outside as soon as the risk of frost is over in early summer.

Right: Almost all squashes are an odd shape. This 'Golden Crookneck' summer squash is typical.

Onion squashes have a fairly sweet orange flesh. Best baked in the oven.

Right: These winter squashes were matured outdoors before being moved into a greenhouse or conservatory. The ripening process is important as they will only keep if the outer skin is hard and disease-free.

The pajama squash has many seeds and is quite watery. It is fairly tasteless.

The pumpkin can be used in sweet or savory dishes.

The 'Ponca Butternut' has few seeds and a sweet, rich nutty flavor. Ideal for baking and can be served as a sweet dish.

Pumpkins have firm golden flesh surrounding a mass of seeds.

Good varieties

Summer squash: 'Custard White'
'Tender and True'.

Winter squash: 'Butternut'
'Vegetable Spaghetti'
'Hubbard's Golden'
'Table Ace'
'Turk's Turban'

Pumpkin: 'Mammoth'
'Atlantic Giant'
'Crown Prince'
'Cinderella'

Right: *Pumpkin 'Cinderella'. These very decorative pumpkins grow slowly to a large size and develop a flattened top and attractive lobes. The rich flesh is edible or they can be used for ornamental purposes.*

'Turk's Turban' has tough skin and the flesh is best boiled and mashed.

'Sweet Dumpling' squash has the same tasty flesh as 'Delicata'. The skins of both can be eaten.

Below: *The 'Atlantic Giant' pumpkin is one of the varieties that can reach over 500lb(230kg) in weight. They need a long growing season, plenty of room and should be restricted to one fruit per plant to achieve this. They are grown mainly for competition.*

Golden acorn squash. The orange flesh has a potato texture.

'Little Gem' squash. The orange flesh is slightly watery. Inedible skin.

'Delicata' has deliciously sweet, chestnut-flavored golden flesh.

Growing courgettes (zucchini)

Twenty years ago, these vegetables attracted only scant attention, yet today gardeners grow courgettes (zucchini) as though they always have. Grow them in a sunny position and in fertile soil well supplied with bulky organic matter. As they are frost-tender, sow them under cover in mid-spring for planting outside in early summer or outside in late spring where they are to grow.

Courgettes were originally just young marrows. Now there are particularly prolific varieties that will carry on fruiting over a long period. Many will continue bearing even if you leave one or two fruits to develop into marrows. As a rule, though, you should pick over the plants regularly and often to ensure a good succession of courgettes. Try not to let them reach more than about 6in(15cm) long.

The best varieties of courgette (zucchini) are of bush habit; they take up much less room than the trailing type and are perfect for most gardens. Like marrows, pumpkins, squashes and cucumbers, they are half-hardy and even the slightest touch of frost will damage or even kill them. This means that in a cool temperate climate you must sow them and grow on the seedlings in a greenhouse or a sunny room until they are large enough to plant outside after the risk of frost is over.

1 Raise courgettes in individual peat pots. Water well and put them in a plastic bag in a warm place to germinate. When they come through, put in full light.

2 Given warmth, the young plants will grow quickly and are ready for potting on or planting out when a good number of roots are growing through the sides of the peat pot.

3 Leave the top of the rootball 0.5in (1.25cm) above the soil. This helps to prevent collar rot fungus disease, as the vulnerable area is able to dry out.

Put two plants in a standard-size growing bag.

4 A good initial watering is vital for these plants. It also helps prevent red spider mites becoming established; they dislike a damp atmosphere.

5 *A few weeks later, the plants are growing away nicely. Ignore the slight difference in foliage; this is a varietal characteristic and not harmful in any way.*

6 *Courgettes grow very successfully in the ground. Lay perforated black plastic around the plants to reduce mud splash and the risk of disease.*

8 *With the right care and plenty of water, you can have courgettes throughout the summer. Always grow a proper courgette variety; a marrow will not do.*

7 *Pick courgettes regularly. Not all varieties continue fruiting if some are allowed to grow into marrows. The two youngsters on the right are aborting.*

New potatoes

Maincrop potatoes are uneconomical in a small garden, but with the help of some cloches or a plastic tunnel, you could be lifting enough new potatoes for a meal when those in the shops are still very expensive. Plant two rows at the most and preferably just one. Even new potatoes take up a great deal of space; an earthed-up row will be at least 18in(45cm) wide. Even without covering the rows, you can still make the crop come earlier by 'chitting' the seed potatoes (causing them to sprout) in a warm room or greenhouse in late winter. Do not worry about keeping the potatoes in the light in the early stages, but once the shoots start to appear, put them in full light to prevent the shoots becoming drawn. When the soil has warmed up sufficiently (look for seedling weeds) in early to mid-spring, plant the tubers about 1in(2.5cm) below the surface and 12-15in(30-38cm) apart. As soon as the shoots appear, be ready to rake some earth over them if a night frost is likely. In any event, earth them up when the shoots are about 6in(15cm) tall so that all but the topmost leaves are buried. It is from the buried stems that the new potatoes grow. Given reasonable weather, you should have an eatable crop in about 16 weeks. You can also grow new potatoes in large plant pots or tubs, as shown here.

1 To 'chit' potatoes, stand them in a warm, preferably dark, place in small pots or on an egg tray, with the eyes (buds) pointing upwards.

Potatoes that have been chitted will grow away much faster than unchitted ones after planting out. Once the eyes have developed into shoots, be very careful not to damage them or knock them off.

2 Used soil-based potting mixture or some old growing-bag mix is fine for potatoes. New compost can lead to rather too much top growth.

3 Fill the pot or tub with the potting mixture until it is one third to half full. Firm it down gently. This tub is about 18in(45cm) in diameter.

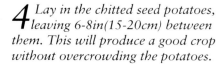

4 Lay in the chitted seed potatoes, leaving 6-8in(15-20cm) between them. This will produce a good crop without overcrowding the potatoes.

5 Cover the potatoes with about 1in(2.5cm) of potting mix and level the surface. There will be enough fertilizer left in the old potting mix.

6 Give the container a thorough watering and stand it in a warm place.

Regular care

The main thing that you must remember when growing potatoes in a sunroom or greenhouse in this way is that the plants are dependent upon you for everything. Make sure that they have enough water and that they do not get too hot. You should not need to feed them, as there will be enough nutrients in the potting mix. If you like, you can move the tub outside, but not until you are sure that there is no risk of frost. The safest time is usually early summer.

9 A fine crop of tasty new potatoes makes all the work worthwhile. Use an early variety; they have the shortest growing season and are therefore ready the soonest. This one is 'Arran Pilot'.

8 It is never easy to be sure, by looking at the tops, when the potatoes are ready for eating. Scrape away the soil to expose a few tubers.

7 A fine display of foliage is the forerunner of a delicious crop. If you have the conditions and can obtain (or save) seed tubers, you can grow potatoes all year round.

Carrots

Carrots are one of the most versatile of vegetables. You can grow them in one form or another throughout the year, from the young pencil carrots that are ready in early summer to the semi-mature roots that can be lifted later in the summer and through the fall. These are followed by the mature vegetables that are stored for use over the winter. Carrots must have a deep and fertile soil that is able to hold plenty of moisture. In particular, you need to grow the early varieties quickly if they are to be tender and tasty. Never grow carrots on land that has just had compost or manure dug in, as they usually produce divided roots. For the earliest crops of tender, young carrots, broadcast the seed, rather than sow it in rows, during early spring. Since the outside temperature will hardly be high enough for germination, use cold frames or lay perforated or slitted clear plastic over the sown seeds as a protection and remove this after about eight weeks.

Good varieties

'Amsterdam': Sow mid- to late winter. Grow under cloches or plastic. Matures early summer.
'Nantes': Sow late winter to mid-spring. Grow in the open. Matures midsummer to early fall.
'Chantenay' and 'Red Cored': Sow mid- to late spring, the crop matures mid- to late fall.
'Autumn King': Sow late spring, matures early winter onwards.

Above: These maincrop carrot seedlings are ready to be thinned and singled to 1in(2.5cm) apart. Thin them again when they touch. The thinnings are delicious to eat.

Growing early carrots

1 Like other vegetable members of the Umberiferae family - parsnip, celery and parsley - carrot seed is slow to germinate.

2 Early carrot varieties may be sown broadcast in cold frames or under clear plastic. Sow thinly, rake the seed in and then water.

3 Early, frame-grown carrots are pulled at this stage. They are sweet and delicate either cooked or sliced into salads.

Below: *These gappy rows of carrots are victims of the carrot fly, whose maggots live in the soil and damage or even kill the vegetable's roots.*

Right: *Because the adult carrot flies stay close to the ground, placing a barrier around the rows will stop them reaching the crop.*

Left: *Here, the mesh surrounding the carrots is clearly visible. Mesh is better than plastic sheet, as it allows the air to circulate freely around the carrots.*

Topped and scrubbed maincrop carrots ready for the pot.

Above: *Juicy carrots make a good winter vegetable. Lift them, clean them and store them in net bags in a dark shed.*

Beetroot

Beetroot is one of those vegetables that you either like or loathe; there seem to be no half-measures. Although they are used a great deal for pickling in vinegar, the freshly boiled, skinned roots make a pleasant and different vegetable all the year round. Beetroot is easy to grow as long as you remember that several of the summer varieties are likely to bolt (run to seed) if they are sown too early in the spring. You can buy bolt-resistant varieties to sow in early spring, but mid- to late spring is more usual and always safer. Bolt-resistant varieties are easily the best to grow and they will provide the main crop in late summer and fall. Further sowings in late spring and early summer can be of long beet that you can store like carrots and use during the winter. The simplest way to grow beetroot is in rows 5-6in(13-15cm) apart. Thin and single the plants to the same distance apart within each row as soon as the seedlings touch. Cover the seedlings with netting to protect them from birds.

Below: *This is how far apart you should grow beetroot. Any closer and the greater competition slows down growth and toughens the beet.*

1 *Beetroot seeds develop in clusters on the plant. In the past, this gave rise to bunched seedlings that had to be singled. Now, seed is available already rubbed (monogerm) from which the seedlings arise singly.*

2 *Using a garden line to be sure of keeping the row straight, take out an even drill 0.8in(2cm) deep in preparation for sowing. You can make the drill with a draw hoe, as shown here, or simply use a stick.*

3 *Sprinkle the seeds thinly and evenly, about an inch (2.5cm) apart. The seedlings will be thinned out later; sowing thinly avoids waste.*

Above: *Be sure to twist the tops off beetroot straight after pulling them. This stops the leaves wilting and absorbing water from the beetroot.*

Below: *This is a fine crop of cylindrical beetroot, but it is still only half-grown. When the vegetables are mature, they can be stored away for use in winter.*

Left: *These beetroot are just the right size for cooking as a fresh vegetable. They could also be pickled in vinegar, but this is usually left for the larger and tougher ones. The single root is a good feature.*

Radishes

Although unrelated to beetroot, radishes also need to be grown quickly so they remain tender with no woodiness. These are summer radishes, but you can also grow winter ones.

Sweetcorn

Sweetcorn is the modern descendant of a very ancient plant native to Central America. Not surprisingly, it is more at home in a Mediterranean climate than in a cool temperate one, but given purpose-bred varieties and a good summer, there is every chance of success in reasonably mild regions. The site should be open and must receive plenty of sun. A soil well supplied with garden compost or manure and sufficient lime to correct strong acidity is also necessary. In order to get good results, sweetcorn must have a long growing season. Modern varieties will be ready for picking about four months after sowing. However, you must balance this with having the plants inside and protected while there is still a risk of spring frosts. Sow the seeds, therefore, in a greenhouse in mid-spring or in a cold frame from mid- to late spring. Plant the seedlings out when the risk of frost is over. If you do not have either a greenhouse or cold frame, you can sow the seeds where the plants are to grow outdoors in late spring or early summer. Once established, keep weeds at bay to avoid competition. When hoeing, be careful not to disturb, and certainly not damage, the surface roots; these are important feeding roots as well as the plant's main support. Provide plenty of water in dry weather. Do not plant traditional and supersweet varieties together, as cross-pollination can affect the quality of the supersweets.

1 The best way of raising young sweetcorn plants is to sow the seeds individually in cardboard tubes filled with seed or multipurpose potting mixture and supported in trays.

2 Place a seed in the top of each tube and press it about 1in(2.5cm) into the mixture. Fill the hole with more mixture and water carefully to avoid washing the seed down.

Sweetcorn varieties

Older varieties, even though many are F1 hybrids, make taller plants than the modern ones and the cobs are coarser and less sweet. These include 'Sunrise', 'Sundance' and the outdated 'Bantam'. The newest varieties are the 'supersweets', so named because of their exceptionally sweet flavor. So sweet are they that you can eat them raw! They also make shorter plants than the older ones. Avoid growing the two types close together as the supersweets are not as good if pollinated by the old varieties.

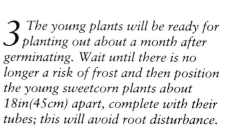

3 The young plants will be ready for planting out about a month after germinating. Wait until there is no longer a risk of frost and then position the young sweetcorn plants about 18in(45cm) apart, complete with their tubes; this will avoid root disturbance.

Left: *Grow sweetcorn plants in blocks rather than rows; this helps the pollination process. These plants are flowering and developing well.*

This is the male part of the plant where the pollen is produced.

These 'silks' are the female part that catch the pollen as it falls from above. Fertilization of the seed follows.

Right: *This semi-mature cob is filling well, but is not quite fat enough to be harvested. The dried up silk is clearly seen; its job completed.*

Right: *Test for maturity by exposing the center of the cob and pushing your thumbnail into a seed. It should be soft but if 'milk' appears, leave it a bit longer.*

Celery

Celery is a vegetable with such a strong and characteristic taste that it is not to everyone's liking. Varieties fall into two distinct groups. The older and traditional kinds are grown in trenches, which are then filled in to blanch (whiten) the stalks. The newer varieties are self-blanching and do not need the trench treatment. Plant these on the surface close together; that is enough to blanch the stems. The soil for celery must be rich in organic matter. This makes it perfectly drained and yet moisture-retentive; just what celery needs. Heavy land is seldom satisfactory unless it is continually improved with bulky organic matter (garden compost, manure, etc.). In addition, slugs are usually troublesome on heavy clay. The best-flavoured celery is still the traditionally trenched sort but modern self-blanching varieties are almost as good and are much easier to grow. Plant these 9-10in (23-25cm) apart in blocks surrounded by wooden planks about 6in(15cm) high to improve the blanching.

Blanching celery

This is a row of trenched celery after its final earthing up. The young plants were set in the trench, enclosed in newspaper, and the trench progressively refilled to its present level.

1 *These self-blanching celery plants were pricked out singly as seedlings. They are now at the right stage for planting out.*

Left: *Late summer and the closeness of the plants is now apparent. This has kept the stalks pale and tender. The blanket mulch was put down before planting to help retain moisture and suppress weeds.*

Right: *Celery as prepared for sale. The leaf stalks are the edible part and you can eat them raw, either on their own or chopped up as part of a salad. Alternatively, you can enjoy them cooked as an individual vegetable or, again, chopped up and used in stews. A very versatile vegetable.*

2 *Water the plants in well; they will need plenty of water throughout the summer months for quick growth and tender stalks.*

Cut off the leaves straight after lifting. If left on, water evaporates through them and the stalks lose their crispness.

51

Tomatoes in the ground

Both single stem and bush varieties of tomatoes will grow perfectly satisfactorily in the ground, either in a greenhouse or outside. First, dig the ground deeply and incorporate plenty of well-rotted garden compost or farmyard manure to provide an abundance of the organic matter that is essential for good results. Raise the plants in the usual way in a greenhouse or indoors on a sunny windowsill. Once they are large enough, greenhouse tomatoes can be planted out at any time in the spring, but do not plant outdoor varieties until the risk of frost is over. Plant bush varieties 20in(50cm) apart, disbudded varieties 2-3in(5-7.5cm) closer, and support the plants with a stout stick to help them withstand wind and rain. Probably the greatest disadvantage of outdoor tomatoes is that the season is a good month (two weeks at each end) shorter than for greenhouse crops. You can partly compensate for this in the fall by covering the ground under the plants with straw, cutting the plants loose from the supporting stake and laying them on the straw. Then place large cloches or plastic tunnels over the plants to keep them warmer and drier, thus extending the season. When even this strategy ceases to work, cut off the fruit trusses and bring them indoors to finish ripening. With luck, they should last until early or midwinter.

1 When the first flowers open, plant out the tomato firmly, but gently. Water it in well and tie it to a cane with soft twine.

2 Unless it is a bush variety, remove any side shoots that appear when plants become established.

3 Remove the side shoot with a sharp tweak. Start from the top of the plant and work downwards.

Sowing tomato seeds

1 Tomato seeds vary in size with the variety. Sow them indoors or in a heated greenhouse.

4 Gently cover each seed with more potting mixture. This is a good way of raising a few plants.

Tomato varieties

Some varieties should only be grown under cover, others outdoors. Remember, they are not interchangeable.

Large-fruited: 'Dombito' (for greenhouse cultivation)
Normal: 'Sonatine' (for greenhouse cultivation) 'Red Alert' (for growing outdoors)
Cherry: 'Sweet 100' (for greenhouse cultivation) 'Gardener's Delight' (for growing outdoors)

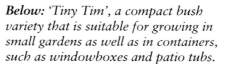

Below: 'Tiny Tim', a compact bush variety that is suitable for growing in small gardens as well as in containers, such as windowboxes and patio tubs.

2 Sow seeds singly in trays of seed or multipurpose potting mix. Put one seed in the center of each cell.

3 Push the seed no more than 0.5in(1.25cm) into the potting mix with the flat end of a pencil.

5 Stand the tray in a shallow basin of water until water soaks up to the surface. Drain for 10 seconds or so.

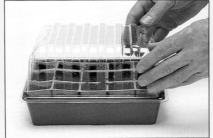

6 Cover the tray with plastic or glass and stand it in a warm place for the seeds to germinate.

Below: This 'Marmande' type of tomato is a single stem variety with huge, ribby fruits. The fruits contain little seed and seed pulp.

Above: Towards the end of the season, plants may show signs of magnesium deficiency, usually brought about by high potash feeds. Not a cause for concern.

Below: A well-grown tomato crop is tasty and can save you money, so it is well worth spending time on it.

Preparing a growing bag

In their simplest form, growing bags are just bags of potting mixture that you cut open and into which you can plant a wide range of both ornamental and edible plants. Growing bags have two main benefits. Firstly, the potting mixture they contain is completely free of the pests and diseases that often affect plants growing in the garden. The other is that they allow you to grow plants in all kinds of confined spaces, including yards, on balconies or windowsills and even on rooftops. There is still a certain amount of suspicion concerning growing bags and a feeling that they are something 'different'. They are not; they are simply another type of container specifically designed for growing plants in. Watering them can sometimes present difficulties, but this applies to watering any plants in containers. The first time you water a growing bag after planting you should apply about a gallon (4 liters) per bag. After that you need not water them again for at least a week unless it is extremely hot; aim to keep the potting mixture evenly damp the whole time. You will find details printed on the bag of how many plants you can grow in one bag. It varies from two marrows, courgettes (zucchini) and melons to three peppers, cucumbers or eggplants (aubergines), six to eight lettuces or twelve strawberry plants.

Opening the bags

There are various methods of opening growing bags. This panel shows one of the three methods featured on these pages.

1 *Another way of preventing an opened bag from spreading too much is to cut off the end beyond the heat seal to form a ring of plastic.*

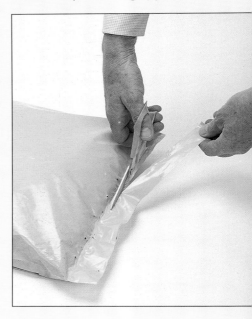

1 *Make a V-shaped cut at one end and cut from the point of the V towards the center of the bag. (Most growing bags have printed guidelines indicating where to cut.)*

2 *Make two planting compartments by leaving a 'bridge' across the middle. The bridge stops the bag from spreading outwards too much when moved or being planted or watered.*

3 *You may prefer simply to cut out the two squares, but folding the edges under makes a stronger job and creates raised sides that prevent water spilling out too easily.*

2 Slip the ring over one end of the unopened bag and maneuver it into the center. Do not be too rough with the strap if it is not very wide.

3 Make a V-shaped cut at each end of the growing bag, with the point of the V towards the center. Join the points of the two Vs with another straight cut running under the strap.

4 Fold under the edges of the single planting compartment and work the mixture down the sides and into the corners of the bag (as in step 5).

The mixtures are formulated to produce the best results for the specific growing conditions. They are not general-purpose mixtures.

5 The potting mixture will have worked its way out of the corners. Push it back and along the sides, so that the surface is slightly dish shaped.

1 If you can prepare and plant up a growing bag in position or if you are going to surround it with, say, ornamental stones, there is less need to prevent it spreading.

2 Cut out a single planting compartment, as shown here. Do not spill the potting mix when refilling the corners.

Tomatoes in a growing bag

One of the best ways of growing tomatoes, in a greenhouse and outdoors, is in a growing bag. Tomatoes are susceptible to root diseases at the best of times and greenhouse plants are especially vulnerable. Growing bags are free of pests and diseases and the plastic isolates plant roots from any diseased soil. Both single stem (disbudded) and bush varieties grow well in bags; raise the plants as described on page 52. Generally speaking, grow one less plant of a bush variety than a disbudded one, because a bush type takes up more room. Follow the general rules for watering growing bags; wait until the surface of the soil has dried out and then give at least a gallon(4 liters) of water at a time. Feeding is not necessary for the first few weeks, but once the first fruits are pea-sized, feed according to the instructions on the bag or the bottle. To allow sun and air to reach the bottom fruits, remove the leaves from the base of the plant up to the lowest truss that has fruit showing red. Tomatoes dislike high growing temperatures, so make sure that the greenhouse is adequately ventilated and shaded. Too much heat leads to excessive water loss and this can cause problems, notably blossom end rot.

3 Firm the plant gently but adequately into place. Never push the soil down too hard or you run the risk of it becoming waterlogged.

1 Never pull a plant from its pot; turn it upside-down, tap the rim on a bench and let the plant drop into your hand. This rootball is full of healthy roots.

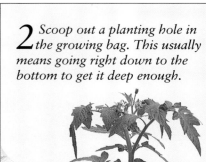

2 Scoop out a planting hole in the growing bag. This usually means going right down to the bottom to get it deep enough.

When to plant out tomatoes

The tomato below is too young to plant out. Extra growth induced by putting it into new soil will lead to an unfruitful bottom truss. The tomato at right has the first flower open and is ready to plant out.

4 *Put in three plants per bag in a greenhouse. Outdoors, plant four single stem, but only three bush plants per bag.*

5 *Apply up to 1.5 gallons(6 liters) of water at the first watering. Later waterings can be slightly less.*

6 *The same growing bag some weeks later. Being a bush variety, the side shoots have not been removed from the plant, so flowers abound with the promise of a good crop.*

Above: *The tomatoes on this bush variety are maturing, providing an attractive array of colors from pale green to bright red. Single stem varieties in growing bags in a greenhouse should grow to six or seven fruit trusses high; outdoors expect to ripen four or five trusses. In both cases, nip out the plant at two leaves above the top truss.*

Growing sweet peppers

Peppers need a Mediterranean climate to flourish outdoors, but in cool, temperate climates they will really only succeed in a greenhouse, whatever you might read to the contrary. Sweet peppers ripen to a bright red, yellow or black, depending on the variety. However, you can use them before they reach their mature color as long as they are fully grown. Sow the seeds in a heated greenhouse or propagator from late winter to mid-spring and prick out the seedlings singly into small pots of multipurpose potting mixture. If you only want a few plants, sow two or three seeds in a small pot and single the plants to one per pot when they are large enough to handle, remove all but the best seedling in each pot. Plant them in their final fruiting position when the first flowers are visible. The plants normally branch out naturally, but if they have not done so by the time they are 12in(30cm) tall, nip out the growing point. Although the peppers are not fully ripe until they reach their mature color, pick the first few while they are still green to encourage more to form. 'Bellboy' and 'Canape' are good varieties. Between planting out peppers and picking the first fruits, their cultivation is much the same as for tomatoes. Give them all the water they require and feed them with a proprietary tomato feed once the first fruits are about the size of a pea. Maintain a moist atmosphere in the greenhouse to discourage red spider mite.

1 Choose a good-quality seed or multipurpose potting mixture and fill as many pots as you will need. Firm and level the surface of the soil.

2 Sow the seeds by spacing them out in a pot or small seed tray. This ensures that seedlings have plenty of room to stop them becoming lanky.

3 Space sowing also makes pricking out easier, as seedling roots do not become jumbled up and need not be pulled apart, which will damage them.

4 Sift some more of the same potting mix and cover the seeds to about their own depth. Never use too fine a sieve as the mix has a tendency to compact and retard germination. Water well.

5 *A stocky young plant, with the first two fruits growing well. If side shoots are reluctant to form, pinch out the tips of the main shoots to encourage them to develop.*

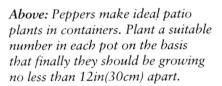

Above: *Peppers make ideal patio plants in containers. Plant a suitable number in each pot on the basis that finally they should be growing no less than 12in(30cm) apart.*

Left: *Peppers come in all shapes and sizes and several colors. These large-fruited ones are ideal for cooking stuffed or just for slicing up as part of a summer salad.*

59

Eggplants (aubergines)

You can raise eggplants (aubergines) in exactly the same way as peppers, but as they require more heat than capsicums, it is better to leave the sowing until early to mid-spring or be prepared to keep them warmer. Do not overdo the heat, though, or they will become drawn, weak and useless. Aim at a temperature in the 60s°F (15-20°C). Never be tempted to take too many fruits from a plant or the size will suffer; four is ample. 'Moneymaker' is a good variety to choose.

When grown in a greenhouse, aubergines are especially susceptible to attacks of red spider mite. Although the plants are fairly well able to stand up for themselves, the mite will spread to other species with disastrous results. There is no reliable chemical control for the microscopic mites, but the predatory mite *Phytoseiulus* produces excellent results if you introduce it as soon as the leaves start to become speckled in the summer. This form of biological control is becoming widely available. In addition to this, you should keep the atmosphere in the greenhouse moist and reasonably cool, which in itself discourages the mite. Whitefly can also be a nuisance but they seldom reach the plague proportions they do with tomatoes. Yellow sticky traps hung above the plants will usually keep them under control, but if they fail, the parasitic *Encarsia* wasp will do the trick.

1 *Sow eggplant seeds singly in small pots For best results, plant them out into a growing bag in a greenhouse or sunroom when they have made a good root system.*

2 *Three plants in a standard-sized growing bag is ample, as they will grow to make large plants. Space the plants out evenly in the bag.*

3 *The top of the rootball should be just below the soil. Firming in the plants excludes air pockets and ensures that water is readily absorbed.*

4 Water thoroughly with at least a gallon (4 liters) of water. This should be all that is needed for 10-14 days. Do not feed at this stage.

5 When the stem is 6-8in(15-20cm) tall, nip out the top to encourage a good number of side shoots to form. It also prevents the plant getting too tall.

6 Nipping out also increases the crop by encouraging more flowers. Support the plants with canes once the weight of fruit bends the branches.

7 These plants have been 'stopped', the side shoots are growing away vigorously and the first flowers are beginning to show.

Once the first fruits are ready for cutting, it is unlikely any new flowers that form will have time to develop fruits.

8 Sooner than you realize, the shiny purple fruits start to form. Although normally needing greenhouse conditions in temperate climates, in a good summer eggplants will succeed outdoors in a sunny, sheltered spot.

Part Two

GROWING FRUIT

In the past, growing fruit in gardens was largely confined to bush and cane fruit, strawberries and, in larger gardens, the occasional apple tree. This was mainly because fruit trees were simply too large for most gardens. Today, however, with the introduction of dwarfing rootstocks and the consequent ability to grow far smaller fruit trees than before, there is no reason at all why you should not grow more or less any fruit that will flourish in the prevailing climate and that takes your fancy. In addition to this, it is easier than ever before to grow good crops of many of the most commonly grown fruits. The main reason for this is that breeding fruit in the last 40 or 50 years has been directed, in part, at creating new varieties that not only crop more heavily and earlier, but that are also more reliable and less susceptible to pests and diseases.

You might say that this has been paid for by a slight deterioration in some characteristics, such as flavor, but this is not as widespread as some like to imagine. The argument that 'you can't beat the old ones' can now be answered with 'oh yes you can', especially when it comes to cropping and reliability. Another thing that encourages some people to grow their own fruit applies equally to vegetables. It is the knowledge that they have only been sprayed with whatever you yourself have applied. All in all, there has never been a better time to start growing your own fruit, or for trying new and perhaps more exotic crops.

Left: Redcurrants sparkling in the sun. Right: A harvest of modern apples.

1 This is a bare-rooted tree just as you would receive it from the nursery in the winter. Note that it has a good root system and strongly growing branches.

Planting and staking a young apple tree

The first thing to remember about planting an apple tree, or any other fruit tree, is that it should·be with you for 20 years or more. This is a long time and you must make sure that the site and soil are absolutely right for it before you start. Any mistakes or omissions now will show themselves in future years. The site for all fruit trees must be sheltered from the wind and yet open to the sun. This is because pollinating insects will only work during the blossom period if the site is warm and free from wind.

Later in the year sunshine will ensure that the fruit develops its full flavor, and shelter from the wind will enable it to remain on·the trees until it is ready for picking. Having chosen the right place for the tree, you must prepare the soil. Allow about 1sq.yd (1sq.m) for each tree. Dig the area and fork the base of the hole to improve the drainage and allow the roots to penetrate deeply. You must also dig in a good quantity of garden compost or manure to make the soil even better. Some recommendations are to add bonemeal to the ground after it has been dug. This is fine if you are planting in the fall or winter but in early spring it is far better to use a general fertilizer Bonemeal only helps new roots to form; a general feed will nourish the whole tree, including the new roots.

2 Dig a hole large enough to take the root system and lay a cane across the top to show how deep to plant. the tree. Drive in a stake before planting so that you do not damage the roots later on.

3 Replace some of the soil in the hole and gently rock the tree about to shake the soil down amongst the roots. Tread it down firmly.

Using a tree tie

1 Nail the buckle end of the tree tie to the stake, pass the free end round the tree and bring it back through the plastic ring.

4 To shape a normal 'bush' tree, select and retain four or five suitable shoots as the future main branches. Start by shortening them by one third to half their length.

Most of the retained shoots are set at a fairly wide angle to the main stem. This gives strong joints that are most unlikely to break under heavy crops.

5 The result of these early stages of pruning is a well-shaped and balanced tree that will produce a strong framework of branches able to carry good crops of healthy apples.

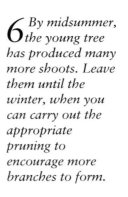

6 By midsummer, the young tree has produced many more shoots. Leave them until the winter, when you can carry out the appropriate pruning to encourage more branches to form.

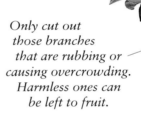

Only cut out those branches that are rubbing or causing overcrowding. Harmless ones can be left to fruit.

2 Push the end of the tie through the buckle so that the tree is held firmly but not too tightly, and away from the stake.

3 Push the tie back through the plastic ring to stop it slipping. If plastic ties are stiff, dip them in warm water.

Left: The same principles apply to planting a tree growing in a container, but remember to remove the bag and any other packaging before you start.

Cordons and espaliers

The cordon is the simplest way of training apple and pear trees intensively. Once you have the principle under control, you can create many interesting and fruitful shapes. Cordons are best planted and trained at a 45° angle to reduce their vigor and encourage fruiting. Cut back any side shoots that are present at planting to three buds straight afterwards. When side shoots appear in subsequent years, prune them in the same way but in late summer. After that, when shoots grow from those already pruned back, shorten the new ones in late summer back to just one bud. Summer pruning encourages fruit buds and fruiting spurs to form. The espalier is really just a vertical cordon but with pairs of horizontal branches 10-12in(25-30cm) apart. Espaliers are trained to wires and can be grown against a wall or in the open. You can form an espalier either by training out a tier of branches to their full length before going up to the next or by starting a fresh tier every year and developing them all together. To grow a tier a year, start with a single-stemmed tree with no side shoots and cut it back to a bud just above and pointing out from the bottom wire. The next two buds down will form the first tier of horizontal branches and the top bud will extend the stem upwards. From then on, do the same every winter until the tree is as tall as you want.

Columnar trees

The new compact columnar (Ballerina) trees, have very close leaves and buds and an almost complete absence of side shoots. They are ideally suited to growing in pots and make first-rate patio trees. Because of their growth habit they require hardly any pruning.

1 *Cordon and espalier apples and pears are best pruned at this stage; in the late summer when they have stopped growing. Winter pruning has less advantages, but will cause no harm.*

Reducing the complexity of the spur improves sap flow to the developing fruitlets.

2 *This spur has already reached a reasonable size. Do not allow it to grow any larger, so cut back the extension growth to a fruit bud. This will increase the fruit size.*

3 *Whether done in late summer or winter, pruning is the same. Cut back new shoots growing from main stems to 3in(7.5cm), and those from previously pruned spurs to 1in(2.5cm).*

Above: *The step-over, or horizontal, cordon is another attractive and space-saving way of growing apples. Grow them beside a path to make them easy to tend.*

Above: *The same tree in full summer growth with a useful crop of apples. It is ready for summer pruning to improve the fruits' size and color. This also reduces pests and diseases.*

4 *By midsummer the criss-cross cordon is growing well and carrying a promising crop. Any fruit thinning must wait until the natural 'drop' is over.*

Family trees

You can buy fruit trees made up of more than one variety of the same fruit. This family apple tree is in full flower and clearly shows the three varieties grafted onto the same rootstock. Ideally, they should all flower together for good pollination, but an overlap of a few days is enough.

'Discovery' - reliable, early dessert apple.

'Fiesta' - a new, late-keeping, Coxlike variety with an excellent flavor.

'Sunset' - prolific midseason with small fruit.

Apple varieties

Once you have decided to grow apples, you must first choose the varieties you are going to plant. Start by making a list of the ones you like. In a small garden, it is far better to grow only those varieties that you already know, as there is simply no room to experiment. Apple varieties are either 'eaters' (dessert) or 'cookers' (culinary), so decide how many of each you would like. As a guide, dessert should normally dominate unless you have a particular liking for cookers. Next, consider the season of use. As a rule, early dessert apples are the most economical to grow because they are the most expensive to buy and do not require storing. This rule applies to any edible crop in the garden; concentrate on those that will save you most if you grow them. Late apples that need to be stored will save money, but you will have to find somewhere to keep them.

Pest- and disease-resistance is very important. Apart from the obviously higher quality of the fruit, the more resistant a variety, the less it will need spraying. The vigor of a variety should also affect your choice. In a small garden it is seldom wise to buy a vigorous variety, such as 'Bramley', as it may soon outgrow its position.

Finally, consider the availability of the varieties you would like to grow. Most garden centers stock very few and these usually include 'Cox' and 'Bramley', which are not the easiest for inexperienced gardeners. A reliable choice, certainly for gardeners in cool temperate areas, would be 'Sunset' (early) and 'Kidd's Orange Red' (midseason) for eaters, with the midseason 'Grenadier' or late-keeping 'Lane's Prince Albert' for a cooker.

Left: 'Egremont Russet' is one of the best russet varieties. It is reliable, has a good nutty flavor, forms fruiting spurs readily and makes an excellent garden apple.

Above: 'Early Victoria' ('Emneth Early') is one of the earliest cooking apples, frequently ready in late summer. It has good flavor, cooks to a fluff and has a compact growth habit.

Left: 'Bramley's Seedling' is still probably the best cooking apple. It often keeps until early spring but because it produces large fruits, it can also be used early in the season.

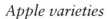

Left: 'Kidd's Orange Red' is one of the best midseason dessert apples, with an excellent flavor and reliable and heavy cropping. Needs a sunny position.

Below: 'Discovery' is one of the finest early eating apples. It has a good flavor if allowed to ripen, and regular, heavy crops with brilliant red color. Do not keep it too long after picking, as it starts to dry out and shrivel.

Right: 'Elstar' is a Dutch apple that is widely grown commercially in Europe. It has a strong and very pleasant flavor, is juicy, sweet and crisp. The color can be weak in a poor summer but it is worth looking for.

Pruning an apple tree in winter

Fruit trees and bushes are pruned to create and maintain their shape, efficiency and health. In the case of apples, you first form the tree, then encourage it to flower and fruit, and finally keep it in good health and growing well. Because we are cutting wood, all the tools must be sharp. The standard pruning tool is a good pair of secateurs. These are used for cutting shoots and small branches no more than 1in(2.5cm) across. There is the scissor type, with crossing blades, and the anvil type, where a blade cuts against a plastic or soft metal pad. The scissor type generally gives a cleaner cut. Loppers are just long-handled secateurs. They are used, not always correctly, for cutting branches too large for secateurs, when a saw is really needed. Cut all branches larger than about 2in(5cm) across with a proper pruning saw, preferably the type with a hollow ground blade.

Always have a reason for cutting out a particular shoot or branch. When using secateurs, cut just above a bud or a side shoot. Both should be pointing in the direction in which you want them to grow. Branches of all sizes must be cut back close to another branch or shoot to aid healing. For the same reason, you should pare smooth all saw cuts. The most straightforward way of pruning bush and half-standard apple trees is by the 'regulated' system, in which all branches that are too high, too low, too spreading or which are growing across the tree are cut out. To these are added any that are dead, badly diseased or broken. Aim to form a tree where every branch is an individual and has room to develop and receive ample sunshine.

Above: Pruning an apple tree encourages it to bear fruit. The first essential is to maintain a balance between growth and fruiting by avoiding congestion.

Above: Here, one branch is crowding the growth of another, so that it cannot perform as it should. Cut out the offending branch. One wise cut will often solve a problem.

This is an overall view of a small apple tree after pruning. Crowded and crossing branches have been cut back or removed, along with any that were too low.

The result is a tidy tree that is easy to look after and cropping well, but not excessively. Nor is it a nuisance to the rest of the garden.

Disease control is another aspect of pruning. The brown shoot shown above is healthy, the silvery gray one below has mildew, so remove it.

Fruit buds

Healthy, plump fruit (blossom) buds will produce the apples. Do not knock them off when pruning. They occur singly or on spurs.

Once a fruiting spur has several buds, cut back any further growth to one.

Right: *This branch is too close to the ground and is likely to interfere with mowing the lawn. Any fruit that it produces will be damaged and useless. Cut it back to a side shoot or spur that will be in nobody's way.*

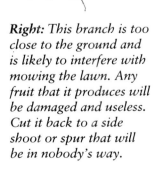

A one-year-old twig has only growth buds. Those towards the tip will develop into shoots during the second year; some or all of the others will develop into fruit buds.

Growth buds

Growth buds are smaller than fruit buds and are flat against the shoots. They are formed at the base of the leaves and usually grow into shoots.

Left: *When cutting back to a growth bud, make an angled cut that is neither too close nor too far away from the bud.*

Right: *A perfectly placed cut allows the bud to grow into a shoot at the correct angle and in the right direction. This is a golden rule of pruning; always cut to a bud so that the wound will heal.*

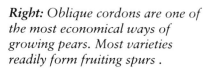

Left: Pears are very amenable to intensive training. This one is trained as a dwarf pyramid and pruned in summer. This results in heavy fruiting on branches of a restricted length.

Pears

Pears are as easy to grow as apples and choice varieties are absolutely delicious. They can be grown in the same tree shapes and sizes as apples, but are even better for growing as cordons and espaliers, because they form fruit buds more readily. You can turn this virtue to your advantage by growing the small, free-standing (not trained) trees called 'dwarf pyramids', which are conical in shape and carry heavy crops. They take up little room and are a good shape for growing in pots and tubs. In management they come between cordons and normal trees. Pears are slightly less easy to assess for ripeness than apples and tend to deteriorate more quickly once they are ripe, so pick pears and ripen them indoors.

Although the choicer varieties are traditionally French and require plenty of summer sun to give of their best, there are some extremely good varieties that succeed well in the cooler summers. A better dessert variety than 'Williams', but slightly later, is 'Onward'. It inherits its juiciness and excellent flavor from 'Comice', one of its parents. 'Comice' ripens later, in mid- to late fall, and is certainly a supreme dessert variety, but it takes a lot of growing to get the best from it. A more reliable alternative is the new 'Concorde'. It ripens at about the same time as 'Conference' and is a 'Comice'/'Conference' cross. The best cooking pear is 'Catillac'.

Above: The espalier, which originated in France, is a series of horizontal cordons growing from a central stem. Again, summer pruning is advisable.

Right: Oblique cordons are one of the most economical ways of growing pears. Most varieties readily form fruiting spurs .

Left: *For many years 'Conference' has been a popular garden and commercial pear. It combines good cropping with an acceptable flavor. The cropping is encouraged by its high fertility.*

Right: *'Comice' (full name 'Doyenne du Comice') is probably the best flavored pear there is. However, it can be a shy cropper and needs at the very least one other variety, and preferably more, planted nearby to ensure good cross-pollination.*

Below: *Although not widely grown in Europe, 'Clapp's Favorite' is a good-quality American, late-summer pear. It is sweet and juicy, if slightly sharp. Eat when it is ready; it soon goes over.*

Right: *'Concorde' is a cross between probably the best flavored pear, 'Comice', and the most reliable one, 'Conference'. It has the heavy cropping of 'Conference' and a much improved flavor through 'Comice'. It ripens in early fall.*

Left: A messy hole on the surface of the fruit signifies codling moth caterpillar attack. Damage is visible from early summer onwards, but at this stage control is impossible.

Cut open a damaged apple and the core area will be black and revolting. You may find the offending pink maggot.

Setting a trap to catch the codling moth

One of the most troublesome pests of apples - and to a lesser extent of pears - is the codling moth. It is the caterpillar of this moth that you often find as a maggot in home-grown apples. The adult moth is a little larger than a clothes moth and lays its eggs on the surface of the fruitlets in late spring and early summer. It is essential that you destroy the caterpillars that hatch from these eggs before they have time to eat into the fruit. You can do this by spraying with, for example, a permethrin insecticide in early summer and again two weeks later. However, a much more convenient and non-chemical method involves attracting and trapping the male moths on a sort of flypaper. This greatly reduces the number of fertile eggs laid and, hence, the number of caterpillars that hatch out. If a trap catches more than about five moths a night or, say, fifteen in a week, then you should spray as well, because you will never catch all the moths. If you catch very few moths, then trapping may be the only control needed. Many garden centers now sell codling moth traps or you can buy them by mail order from a number of sources. There is also a trap available for the red plum maggot (plum fruit moth), an equally serious pest. Both traps are 100% specific and will only catch the one kind of moth.

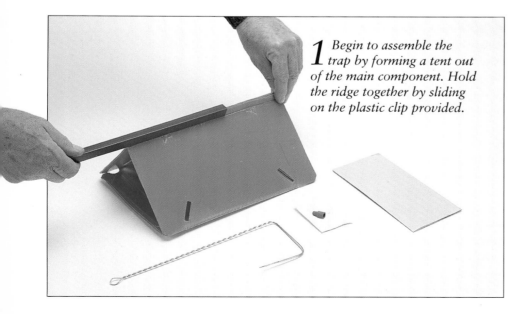

1 Begin to assemble the trap by forming a tent out of the main component. Hold the ridge together by sliding on the plastic clip provided.

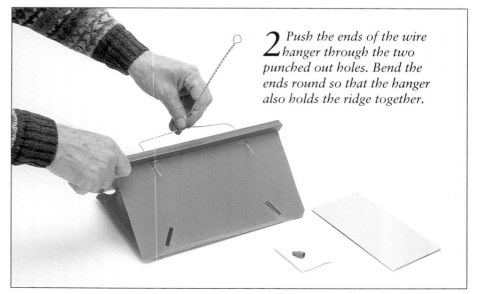

2 Push the ends of the wire hanger through the two punched out holes. Bend the ends round so that the hanger also holds the ridge together.

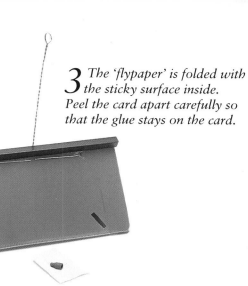

3 The 'flypaper' is folded with the sticky surface inside. Peel the card apart carefully so that the glue stays on the card.

4 A piece of rubber impregnated with the female pheromone (scent attractant) is the bait. Place it in the center of the sticky card.

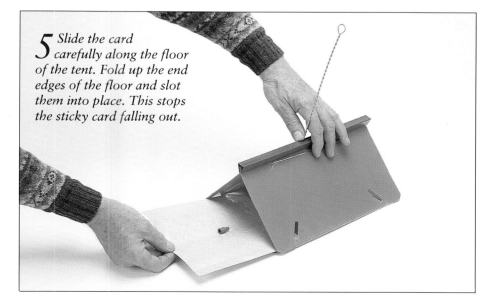

5 Slide the card carefully along the floor of the tent. Fold up the end edges of the floor and slot them into place. This stops the sticky card falling out.

6 Bend the top of the wire hanger into a hook and hang the trap at head height in the apple or pear tree. One trap attracts male moths within about a 30ft(9m) radius. If trees are further away, provide more traps.

7 This trap has more than earned its cost. The pheromone bait will need renewing after about five weeks if it is still attracting moths.

A good catch of male moths, but it indicates just how many there are to damage the crop.

Storing fruit

Only apples and pears can be stored as they are; all other fruits must first be prepared and then frozen, bottled, etc. Late-ripening dessert apples and cooking apples destined for storing can be picked all at once. Pick dessert pears when they are still hard and allow them to ripen indoors. Cooking pears are picked and cooked or stored when hard. When dessert pears are ready for picking, they part quite readily from the tree.

When picking any fruit, but particularly if it is to be stored, treat it gently, as any damage before, during or after picking will ruin it for storage. Only store fruit that is completely sound. For the longest storage, keep apples and pears just a few degrees above freezing. Choose a suitable brick shed or garage and make sure that it contains no material that is likely to taint the fruit. Look at the fruit regularly and remove any that are bad or ready to eat. Apples can be either wrapped and boxed, put in cell-pack boxes or laid out on shelves or racks. Always put pears on racks, as they need air.

1 *A good way of keeping apples is in boxes of cell-packs that you will often find being thrown out by shops and other fruit retailers.*

2 *Place one apple in each cell of the 'concertina'. This keeps them apart, so that any rot will not pass from one fruit to the next.*

3 *Put a dividing sheet of paper in between each layer of fruit. This not only helps to keep the fruits apart, but also makes the pack more stable.*

4 *Carry on filling the box with fruit until it is full. Always keep just one variety in a box; different varieties seldom ripen at the same time.*

The amount of russet is immaterial, but the green ground color shows that this 'Conference' pear is not yet ripe.

This fruit is ready for eating. The green has turned yellow and the flesh will give when pressed.

Apples are best left with the natural bloom on when stored and certainly on the show bench.

'Spartan' is naturally shiny and often darker than these. A good shine makes it more appetizing.

Left: *Another good way of storing apples is to wrap them individually in kitchen paper. As with cell-packed fruits, this prevents the spread of any storage rot that may appear after a while.*

Below: *Storing apples and pears on racks is possibly the best way, but also takes up the most room. Air can circulate, and ripe or rotting fruit is easy to see.*

5 Once the box is full, lay a piece of card or a folded newspaper on top and put it away in a cool, dark and airy place to store the apples.

Above: *This is a homemade model that has the added advantage of being mouse-proof by virtue of the wire netting back and front.*

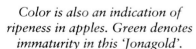

Color is also an indication of ripeness in apples. Green denotes immaturity in this 'Jonagold'.

Once the ground color changes to yellow, bring the fruit into a warm room to finish ripening.

Exactly the same ripeness indications are seen in a wholly green variety when unripe.

When ripe, this 'Crispin' mellows to yellow, with a blush on the cheek.

Planting a peach tree or nectarine

Essentially, you grow and treat peaches and nectarines in exactly the same way; the nectarine was originally a hairless 'sport' found growing on a peach tree. Even today, you occasionally find nectarines on peach trees and vice versa. Everything, therefore, that is said about growing peaches refers equally to nectarines. They come from the Mediterranean and, as such, are not completely at home or hardy in cooler climates, so give them the warmest position available in the garden. They also flower early in spring and you will normally need to hand pollinate them because few pollinating insects have emerged from hibernation when the blossom is out. Hand pollination of the flowers is very easy. Wait for a warm, sunny day when the flowers are open. Using an artist's soft paintbrush or even a piece of absorbent cotton, draw it over the open flower. This collects pollen from the anthers and transfers some to the receptive stigma. Given relatively warm weather, fertilization will follow and a fruit will develop. If there is a good set of fruit you will probably need to thin the fruitlets to ensure that fruit size does not suffer. Carry out a first thinning in early summer when the fruitlets are the size of a hazelnut, so that they are 4in(10cm) apart. Follow this with a second thinning two to three weeks later, when they have reached walnut size. Aim to leave the fruitlets about 8in(20cm) apart. Reliable varieties of peach for outdoors include 'Peregrine' and 'Rochester'. 'Early Rivers' and 'Lord Napier' are good nectarines.

1 A young peach tree straight from the garden center. When planting it, use the rootball to judge the size of hole you need to dig.

2 Once firmly planted, the tree should be very slightly deeper than it was in its original pot, so that new roots form from the stock.

3 Now comes the hard part. Cut back the tree to leave two suitable shoots to form the first rays of the fan. This treatment means that you can start training from the right basis.

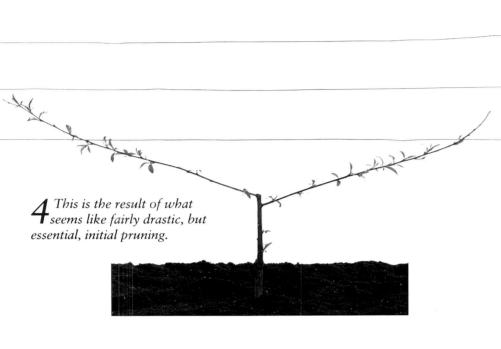

4 This is the result of what seems like fairly drastic, but essential, initial pruning.

5 Cut back these two shoots to two-thirds to threequarters of their original length. Prune to a bud on the upper surface of the shoot so that the resulting growth grows upwards.

6 Tie two primary training canes to permanent wires, 9-12in(23-30cm) apart. Use plastic string, as it lasts longer. Wire is even better, but bend the sharp ends out of harm's way.

7 This is how the tree and canes will look when they are all in position. The ends of the canes are close to the tree.

8 Using soft string, tie the shoots to the canes, but not too tightly. This avoids strangling the shoots as they grow, which can happen with plastic.

9 The tree is on the point of starting to send out new shoots from the two that were trained in. If necessary, protect the tender growths from frost.

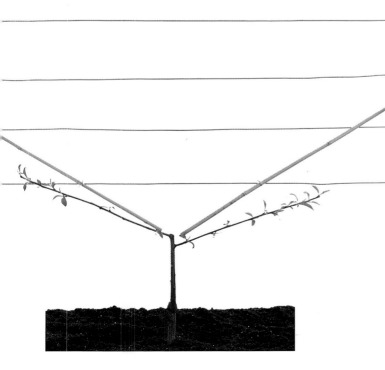

Tying the shoots to a cane helps them to grow straight and lends support when they bear fruit.

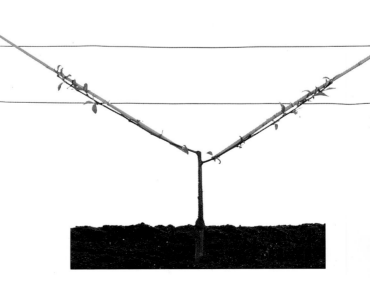

How a peach tree fruits

This young peach tree is the same age as the one we have been following through its first year after planting. Peach trees only flower on the shoots that grew during the previous growing season. This is clearly seen by the blossom on the one-year-old shoots.

Training a peach tree

Being natives of the Mediterranean, peaches and nectarines can seldom be grown as free-standing trees in the open garden in cooler countries. To give of their best they should be fan-trained against a warm, sunny wall or fence. Given the choice, a wall is better than a fence because the brickwork holds a great deal more heat. This is released at night and will frequently raise the temperature around the tree sufficiently to keep away a slight frost. The sequence of training a tree in a fan shape started on page 78 continues here. Try to follow the guidelines as closely as possible because what is done in the early years has a big influence on the tree's future.

Unquestionably, the worst affliction to attack a peach or nectarine tree is a disfiguring fungus disease called peach leaf curl. This causes first the young leaves and then the older ones to become blistered, swollen and bright red. Control is difficult, but covering the fan-trained trees with plastic when it rains during the growing season keeps the leaves dry and greatly reduces the spread of this devastating fungus. Another approach is to make two applications of a copper spray, a week apart, at leaf-fall and another two when the buds have started to grow out in spring. These will go a long way to control the fungus.

1 *By midsummer the tree is growing well. The right side is more vigorous than the left; correct this later by appropriate pruning.*

2 *Fix another cane in position for a new shoot. Tie it temporarily to the stem of the tree to help training and release it at a later stage.*

3 *Tie the shoot that has developed from the lower surface of the main branch to the new cane. This will become the lowest branch.*

4 *Fix another cane for the shoot growing from the upper surface of the branch. Pinch off any shoots that are pointing away from the tree.*

5 *As before, tie in all the new shoots with soft string, but do so quite loosely to avoid any constriction of the shoot. Examine all ties regularly.*

Above: *The reward for attentive management; a crop of delicious peaches in the making.*

The bareness on this side of the tree can be overcome by appropriate pruning and by tying the weak branch into a more upright position until it grows more strongly.

6 *Although it is still summer, the tree has almost finished its first growing season. There will be little further growth from now on.*

Plums, gages and damsons

Plums of all sorts are becoming more popular these days. The smaller trees grown on the rootstock 'Pixy' are much easier to manage, and you can grow plum trees as cordons and fans trained against walls and fences, 'festooned' trees with their branches tied down to encourage fruiting and also pyramid trees. Spring frosts are still a problem, but with the smaller, more compact trees simply drape some horticultural fleece over them on spring evenings when a frost is likely to strike. However, you should still grow plums in a sheltered and sunny part of the garden because this helps the fruit to ripen. In its early years, concentrate on forming the tree and building up its branches by appropriate pruning. Later, though, ordinary bush trees require very little attention beyond the removal of shoots and branches that are dead or diseased, crossing or crowding, or that are growing too high, too low or are too spreading.

There are very few plum diseases to bother you. The worst is the silver leaf fungus in which the normal green of the leaves develops a metallic sheen. The best way of preventing this is to prune when the tree is in leaf, because the fungus only spreads in the winter and by entering fresh wounds. The red plum maggot is also tiresome, but it can now be controlled with a specific pheromone trap, similar to the codling moth (see pages 74-75). 'Victoria' is still the most reliable variety for gardens. A good early cooker is 'Czar', while 'Kirke's Blue' and 'Cambridge Gage' are better eaters than 'Victoria'. The 'Prune' or 'Shropshire' damson has the best flavor but 'Farleigh' the heaviest crops.

1 This tree has already been trained in the nursery as a fan. Four branches are all that are needed so the rest are removed with clean secateurs.

2 Cut close to the topmost shoot that is to be retained so that the wound will quickly heal over and exclude any harmful fungal rots.

Left: *Although not often seen in gardens, 'Marjorie's Seedling' is a reliable cooker. The tree is tall but flowers and fruits late, so is useful in frosty areas.*

Below: *Damsons are small cooking plums that often crop prolifically. 'Bradley's King' crops in early fall. One of the largest and best-flavored damsons.*

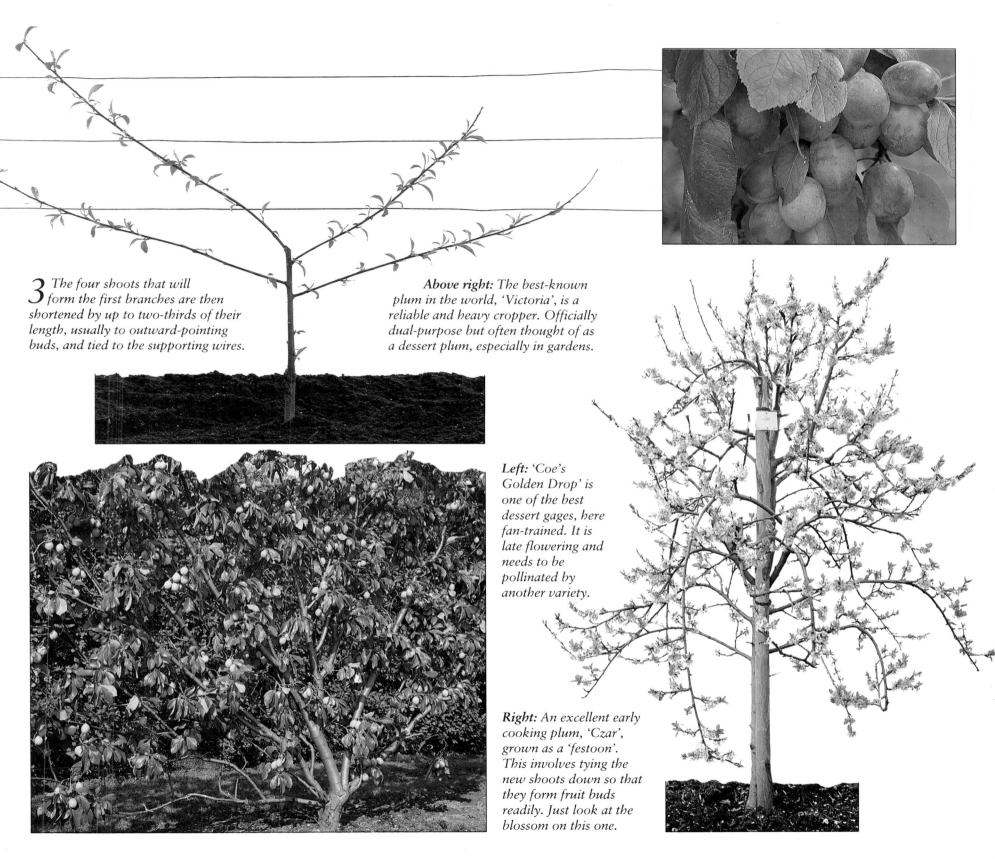

3 The four shoots that will form the first branches are then shortened by up to two-thirds of their length, usually to outward-pointing buds, and tied to the supporting wires.

Above right: The best-known plum in the world, 'Victoria', is a reliable and heavy cropper. Officially dual-purpose but often thought of as a dessert plum, especially in gardens.

Left: 'Coe's Golden Drop' is one of the best dessert gages, here fan-trained. It is late flowering and needs to be pollinated by another variety.

Right: An excellent early cooking plum, 'Czar', grown as a 'festoon'. This involves tying the new shoots down so that they form fruit buds readily. Just look at the blossom on this one.

1 Before planting a young gooseberry bush such as this, remove any shoots that are closer than about 5in(13cm) to the roots, as well as any buds among the roots.

Growing gooseberries

Gooseberries need good soil conditions to give of their best. This is especially true of dessert varieties, which are grown to a larger size than cookers. Take care when tending bushes or picking fruit, as the thorns are long, pointed and very sharp. One of the best ways of growing gooseberries is as cordons trained to a wall or fence or grow them in the open up canes. They are much easier to look after than bushes, the dessert varieties ripen better and you are less likely to be scratched. You can grow cordons with just one vertical shoot, or double or multiple cordons with two or more shoots. Single cordons are the quickest to reach the desired height but you need many more plants for a given length of fence. Another unusual shape is the standard, which, like a standard rose, is just a bush on top of a 3-4ft(90-120cm) stem. Ordinary bushes take up rather a lot of room and are not as easy to maintain. 'Careless' or 'Jubilee Careless' are the most prolific cookers, with 'Leveller' an excellent all-round dessert variety.

2 Plant the bush firmly, leaving a clear leg of 5-6in(13-15cm) between the ground and the first branch. Nick out any buds on the leg.

3 After planting, prune the bush by removing any weak shoots, cutting back misplaced ones and shortening the rest by roughly half their length.

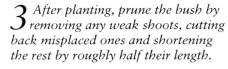

Taking gooseberry cuttings

Gooseberries are easy to propagate. Simply take hardwood cuttings, retaining just the top three or four buds, and insert them into V-slits in the ground 6in(15cm) apart. Push the cuttings into the ground as near upright as possible so that the resulting bushes have a vertical 'leg'. The best time to take cuttings is in early winter as soon as the leaves have fallen.

Left: 'Whitesmith' is a first-rate dessert gooseberry. It crops well and has an excellent flavor. The medium vigor of the bush makes it very suitable for growing as a cordon.

Below: A row of well-trained and maintained 'U' cordons cropping well after three or four years growth. Net the fruit to protect it from birds.

4 Pruning stimulates side shoots, so that the original four or five 'branches' give rise to eight to ten after the first growing season. This forms the framework of the bush.

5 For a U-cordon (with two upright stems), prune back the bush to two strong shoots growing out opposite each other.

Bend the two shoots down carefully and tie them as horizontally as possible.

Pruning gooseberries

For the biggest and best dessert gooseberries, prune both cordons and bushes in early summer. That is also the time to thin out the fruits to one per cluster, using the thinnings for cooking. Summer pruning simply involves cutting back the new shoots (those you do not want for extension growth) to five leaves, about 5in(13cm). Summer pruning improves the berry size and color of dessert varieties and the crop weight of cookers. Early to midsummer is the normal and the most beneficial time to summer prune. Because you are removing the soft growth at the end of the shoots, which is the part most susceptible to attack by mildew, summer pruning also plays a very valuable role in controlling this disease. As well as this, you should also spray with benomyl or carbendazim. As far as pests are concerned, the gooseberry sawfly is the worst. Control it by spraying with permethrin or derris in mid-spring. Unfortunately, there are no thornless varieties worth growing, but cordons are a great deal less painful to look after than normal bushes.

1 A mature, unpruned gooseberry bush in winter has shoots and branches growing in all directions and everything is covered in thorns.

2 Remove all shoots and branches growing too low or too far out. Crossing branches and those crowding others must also go.

3 Once the framework is tamed and tidied up, begin shortening all the branch leaders (young shoots on the end of each branch) to half their length.

This is a strong, young branch that will carry heavy crops.

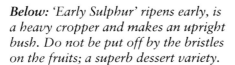

Below: 'Early Sulphur' ripens early, is a heavy cropper and makes an upright bush. Do not be put off by the bristles on the fruits; a superb dessert variety.

Right: 'Lancashire Lad' is a large-berried, coarsely bristled but popular dessert variety. It ripens in early to midseason and has a good flavor.

Above: 'Leveller' is a heavy cropper, with good flavor that ripens mid- to late season. The main commercial dessert variety.

4 Shorten side shoots on existing spurs or growing straight from main branches to 3in(7.5cm) to an up- or outward-pointing bud.

Completely remove any weak shoots.

5 The tidily pruned bush is well equipped to carry heavy crops without collapsing under the strain. Upright training counteracts the gooseberry's natural weeping habit.

Planting blackcurrants

1 *Make sure that you start with a good-quality, strong growing bush. This one-year-old example has three vigorous shoots and a good root system.*

Blackcurrants are an excellent soft fruit for making into jellies and fruit fools, for filling pies and fruit tarts and they make a novel addition to summer fruit salads. They are also exceptionally rich in vitamin C and can be used for cordial or wine-making. Unfortunately, traditional bushes of most varieties are too large for many gardens, but by choosing a more compact variety and pruning it harder than normal, they become a more viable proposition. Nor is it possible to grow blackcurrants as cordons, because their best fruits and heaviest crops are produced on young shoots, which are removed when pruning cordons. Blackcurrants are very easy to propagate using hardwood cuttings taken in the early winter. You can start as soon as the leaves drop in late fall.

The worst pest is the blackcurrant gall mite. Not only does it cause the buds to become swollen and stay closed in spring, but it also spreads the virus disease reversion. There is no cure for either condition, so be sure to pull up and burn any bushes bearing even just a few of the unopened, pea-sized buds in spring. Both the mites and reversion are confined to blackcurrants.

2 *Dig out a hole wide enough to take the root system and deep enough to allow the bush to be planted about 2in(5cm) deeper than it was before.*

Spread out the roots in the planting hole.

4 *After planting at the correct depth (shown here) completely remove all the weak shoots and shorten the remainder to two or three buds long.*

3 *Cover with soil. Shake the bush so that each root is in contact with the earth. Firm down soil to prevent it drying out and to support the bush.*

Taking cuttings

1 *Push a spade into the ground 8-9in(20-23cm) deep and lever back on the handle to form a slit to receive the prepared cuttings.*

2 *Prepare cuttings in midwinter from strong, straight, healthy one-year-old shoots. Carefully cut back thin tips to a strong bud.*

3 *Cut the base of the shoot to a strong bud. The cutting should be straight and strong and 9-12in(23-30cm) long.*

4 *Push cuttings into the slit 6in (15cm) apart, so that only the top buds are visible. Strong shoots will come from below ground.*

5 *Push the soil down around the cuttings really well so that they are firmly in place for the winter. There is no need to water them in.*

6 *These cuttings have two buds showing above the surface, but do not worry if there is only one.*

7 *Here is the same bush the following summer. There are at least three new strong shoots. These will not be pruned and will start fruiting the following summer.*

Generally speaking, you should prune back to an outward-pointing bud.

5 *This harsh treatment encourages strong shoots to grow from the base and stimulates the root system, so that the bush establishes quickly.*

6 *Always cut back to a bud that in the following growing season will give rise to a new shoot developing in the desired direction.*

Pruning blackcurrants

The aim of pruning is to keep the bushes young and as compact as possible. It helps to choose the least vigorous variety, the relatively new 'Ben Sarek', with a recommended planting distance of 36in(90cm), instead of the more normal 48in(1.2m). Blackcurrants are usually pruned in early winter by cutting out branch systems when they reach four years old. If you want to keep the bushes small, reduce this to three or even two years. Because the bushes will be smaller, also reduce the planting distance between them in the first place or the crop from a given length of row will be considerably lighter. A simple way to establish the age of a branch is to start with the young shoot at the tip and count backwards down the branch. Each year's growth is darker. As well as 'Ben Sarek', other good garden varieties include 'Ben Lomond' and 'Ben Connan'.
This is a new one that is reputed to have the largest individual currants of all and to have good disease-resistance, too.

Right: Jostaberry, a vigorous hybrid between blackcurrant and gooseberry, crops heavily and may be trained and pruned into many shapes and sizes.

1 Even a well-tended blackcurrant bush can seem daunting when you come to prune it. In fact, you need only follow a few simple rules.

2 Cut out all branches growing too close to the ground, across the bush or that are more than three years old. Eliminate any that cause congestion.

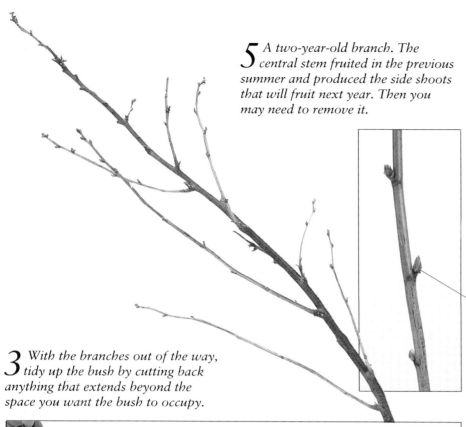

5 A two-year-old branch. The central stem fruited in the previous summer and produced the side shoots that will fruit next year. Then you may need to remove it.

3 With the branches out of the way, tidy up the bush by cutting back anything that extends beyond the space you want the bush to occupy.

Right: 'Ben Sarek' is a relative newcomer among blackcurrants. The bush is significantly smaller than other varieties, yet it is heavy cropping. An excellent garden variety.

A one-year-old blackcurrant shoot. Note the fat fruit buds that will carry fruit next summer. Do not remove a shoot like this.

4 Virtually no shoots have been shortened. If one is too long, cut it right out. Shortening a young shoot encourages growth and branching, which makes it even larger.

Redcurrants and whitecurrants

This shoot arises too close to the base of the stem and needs cutting out straight away.

Cultivated not as conventional bushes but as cordons, both redcurrants, and their colorless cousins the whitecurrants, take up very little room and will grow either in the open garden trained to canes or against a wall or fence. U-cordons are more economical of plants than single vertical cordons, but the latter will cover the area more quickly. Pruning is best carried out in early summer by cutting back the new shoots to within 4in(10cm) of the main branches and, further, to 2in(5cm) in the winter. Summer pruning encourages the currants to grow and ripen and also reduces the amount of soft, sappy growth, which would be susceptible to mildew disease. The Dutch variety 'Jonkheer van Tets' (early) and 'Red Lake' (midseason) are both excellent in gardens. Grow 'Stanza' for a late crop. 'White Versailles' is the most widely grown whitecurrant. Propagate both kinds by hardwood cuttings in early winter. Although mildew is sometimes seen, it is never as serious as it is on gooseberries. The worst pests are usually aphids that cause red blisters to form on leaves. If this is serious, any systemic insecticide will control them.

This young and well positioned shoot will make a fruitful branch.

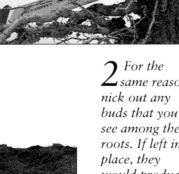

Shorten strong shoots to an outwardly pointing bud.

2 *For the same reason, nick out any buds that you see among the roots. If left in place, they would produce unwanted suckers later on.*

1 Start with a strong, one-year-old bush. As it is best grown on a short 'leg', remove any shoots that are less than about 5in(13cm) from the roots.

3 Make the planting hole wide enough to take the spread-out root system and deep enough to leave the planted bush just deeper than before.

4 After planting, remove weak shoots. Cut back strong but badly placed shoots to 1in(2.5cm). Shorten remaining shoots by half their length.

2 *Shorten the branch leaders (the year-old top section) by half and cut back unwanted side shoots to about 2in(5cm).*

A cluster of fruit buds. Pruning will encourage more buds to form. Do not cut these buds off.

Taking cuttings

Propagating redcurrants and whitecurrants is largely the same as for blackcurrants (page 89). Only retain the top four buds, however, and push in the cutting to leave 6in(15cm) between the lowest retained bud and the ground. Insert 6in(15cm) apart.

1 *To prune an established bush, remove any low, broken or crossing branches, plus any that are clearly causing overcrowding.*

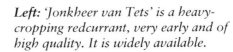

Left: *White-currants are more of a novelty. The crops are not as heavy, but the flavor is similar to redcurrants. They require the same cultivation.*

Left: *'Jonkheer van Tets' is a heavy-cropping redcurrant, very early and of high quality. It is widely available.*

Right: *The pink currant makes an interesting addition to a fruit garden, and has the same qualities as the reds.*

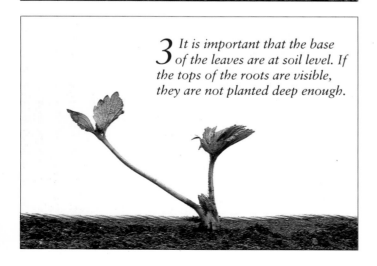

1 *Buy strong, bare-rooted plants in late summer or early fall. This one was propagated just a few weeks previously. Do not be tempted to choose larger but older plants growing in pots.*

2 *Use a garden line to mark the row and make a hole deep enough to take the whole root system without having to bend the roots to fit them in.*

3 *It is important that the base of the leaves are at soil level. If the tops of the roots are visible, they are not planted deep enough.*

Planting strawberries

Strawberries are probably the most popular and easily grown fruit, and you can include some plants even in the smallest garden. As well as the traditional rows of plants in the vegetable section, you can also grow strawberries in pots, tower pots, growing-bags, hanging baskets, windowboxes and other planters or even as edging plants in the flower garden. You can use the fruits in many ways, to make desserts, jellies and in all sorts of puddings. They are truly a versatile and delicious fruit.

Strawberry varieties are either of the summer-fruiting or perpetual-fruiting type. Summer-fruiting are the most popular and you can advance them under cloches or even in an unheated greenhouse. For later crops, use one of the perpetual varieties. These will start flowering and fruiting naturally at about the same time as the summer ones, but by removing the flowers as they appear until very early summer, you can induce them to fruit from late summer often until late fall, depending on the weather.

The worst disease is unquestionably botrytis (gray mold), which turns the fruits into gray puffballs of fungus. You can reduce the infection by putting down straw between the rows or using a proprietary alternative to keep mud off the fruit. In addition, spray fortnightly with benomyl once the first flowers are showing color. Strawberries will nearly always need netting against birds. If you want to keep summer-fruiting varieties for another year, cut off the leaves, old fruit stalks and unwanted runners after fruiting. This cleans away many pests and diseases and rejuvenates the plants. This 'haircut' is not essential but it is certainly beneficial. If you give them a high-potash feed at the same time, they will build up into strong plants for the winter and following year.

Below: *Delicious crops of strawberries like this are perfectly possible if you take care over them and grow modern varieties, such as 'Elsanta'.*

1 Once the baby strawberries are about the size of a pea, you must cover the ground under the plants with straw or mats.

2 Open the mat and slip it carefully around the base of the plant. You can make your own protective mats from old carpet or underlay.

This young plant in its first cropping year fits the proprietary mat well but it may be too large for it next year. Beware of this.

Above: *Netting to protect the crop is vital. This system is splendid; it saves the strawberries and looks attractive.*

Left: *Strawberries this large are often utterly tasteless. Not so 'Maxim'. However, this size of fruit is only found in the first crop. After that it returns to a more normal size.*

3 Work plenty of clean straw well under the plants so that the fruitlets are not splashed with rain and mud, which usually leads to botrytis.

Propagating strawberries

1 *Propagate strawberries from the plantlets that form at the end of 'runners'. These grow from the main plant in the summer.*

2 *Fill some pots with used potting mixture and sink them in the ground around the plant whose plantlets will be used for propagation.*

The number of runners and plantlets produced will vary with the variety. Never use more than five per plant or they are likely to be weak.

Because new plantlets are naturally produced on 'runners' sent out by established plants, propagating from your own strawberries is so easy that it is sometimes done without a thought for the snags involved. The main - and very important one - is that you could be using plants that are infected with a virus disease. Clearly that is a recipe for disaster, as the new plants will also be infected and will never crop well. There is no doubt that if you are uncertain about the health of your plants, the only sensible action is to buy new ones that carry a health certificate from a nursery or garden center. However, assuming that all is well, the first thing to do is to select what you consider to be the best parent plants. They should be strong-growing, free from pests and diseases, and good croppers. Make the selection during the early part of the growing season and into the fruiting period, so that you can start propagating straight after the plants have fruited in early to midsummer. It is also a good idea to work out beforehand how many new plants you will need. On the basis of rooting no more than five plantlets per parent plant, you will then know how many plants you need to nominate as parent plants.

The simplest way to propagate strawberries is to peg down the plantlets (one per runner and leaving them attached) onto the soil surface near the parent. However, it can be inconvenient having the new plants dotted about all over the place, so a much better plan is to peg them into 7-9in(18-23cm) pots of old potting mixture. Not only is rooting quicker into the more friendly medium, but you can also move the rooted plants after they have been parted from the parent a month later.

3 *Leaving it attached to the parent plant, peg the plantlet down into the potting mixture with a piece of looped wire. This will prevent any movement during rooting.*

Below: *Although you could use all three plantlets on this runner, it is better to choose just the strongest one.*

4 *The pot may be watered after 'layering' the plantlet to settle the potting mixture, but that is usually the only time any watering is needed.*

This is the runner from the parent. The plantlet grew from the end of it.

5 *After about a month, a good root system will have developed in the pot and you can cut the new plant from the parent.*

6 *Remove the remains of the runner and any dead leaves and flower stalks. If planted by early fall, you can expect a full crop the next summer.*

7 *Dig a hole large enough to hold the rootball comfortably. Make sure it is sufficiently deep to allow the base of the leaves to sit at soil level after planting.*

8 *Push the soil back around the rootball and firm it in well to exclude all the air pockets. This leads to quick rooting.*

9 *Water the plant in to settle it into its new planting position. It now has every chance of producing a heavy crop of fruit the following summer.*

Raspberries

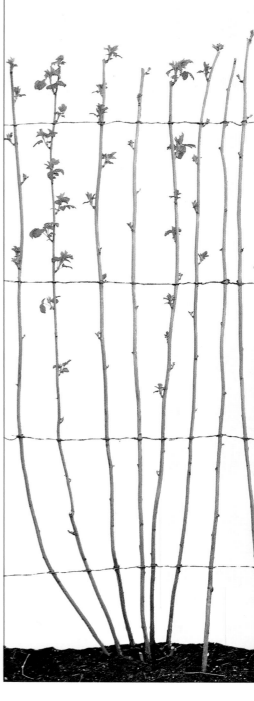

Raspberries are a popular and easy-to-grow soft fruit. The most convenient way to grow them is in rows in the vegetable garden, although by planting them in groups of three it is quite possible to grow just a few in mixed borders or elsewhere tied to stakes. There are two types of raspberry: summer-fruiting and fall-fruiting. Apart from the obvious difference, the main one is that the fall varieties produce canes that fruit later in the same season, whereas the canes of summer varieties grow during the first year and fruit in the summer of the second year.

Each cane carries just one crop of fruit and the pruning strategy is based on this. With summer varieties, cut down the fruited canes straight after fruiting and tie in the new ones to the wires in their place. All the canes of the fall varieties fruit together. Once they have fruited, leave them until new ones start to appear above ground in the following spring and then cut all the previous year's canes to the ground.

Left: Good, bare-root summer-fruiting raspberry canes are available during the late fall or in winter. Avoid containerized plants, as they are far more expensive and no better.

1 *Plant the canes 15in(38cm) apart, so that all the roots and any buds on them are below the surface. These buds will grow out to produce the following year's canes.*

2 *Immediately after planting, cut back the canes leaving them about 10in(25cm) long. This will allow just a few new shoots to grow from them in the following year.*

Some fruit will develop on the side shoots next year. Only allow a very few to grow on.

3 *It is important to have these new shoots on the canes in the next growing season because they will keep the canes alive while more are being sent up from the buds on the roots.*

Above: In early spring, the canes from these well-established plants are starting to send out the side shoots that will carry the fruit. Notice the ample spacing and supporting wires.

Fall raspberries

The fall-fruiting raspberries extend the fresh fruit season well into the fall - until the first frosts, in fact. They crop as heavily as the summer varieties, but over a longer period. As well as fruiting on the new canes, another difference is that they are much more vigorous. You must, therefore, keep the width of a row to no more than 18in(45cm) from one side to the other.

Left: Fall raspberries need no permanent support. Pass some twine down both sides of the row and secure it at each end.

Right: 'Autumn Bliss' is a superb fall variety and provides fresh fruit until the first frosts put a stop to cropping.

Below: A good stand of strong, new canes after pruning. Tie them to the supporting wires with soft twine to leave them about 4in(10cm) apart.

Above: The early and popular summer raspberry 'Glen Clova'. A regular and heavy cropper, with medium-sized berries of good flavor.

Above: If, after tying in, the tops of the new canes reach well above the top wire, bend them over carefully and tie them down to prevent damage.

Below: Prune summer raspberries straight after fruiting. Cut out all the canes that have fruited, along with any new ones that appear weak.

Blackberries and hybrid cane fruits

Since the introduction of improved thornless varieties, blackberries have become more popular as a garden fruit. There are many ways of training blackberries, but with the exception of 'Loch Ness', all methods involve tying the canes to horizontal wires so that they can be tended and picked easily. One feature common to all systems is that the new canes are kept separate - and usually above - those that are about to fruit to reduce the likelihood of diseases being passed from old to new canes. The less vigorous 'Loch Ness' need not be tied to wires; simply train it onto vertical poles about 8ft(2.5m) tall. It is a good variety for small gardens.

With very few exceptions, hybrid cane fruits are all hybrids between blackberries and raspberries. Mostly, they have the long, supple canes of the blackberry but, except for the few thornless variants, their numerous small thorns are more like those of the raspberry. Training systems are similar to those for blackberries. Despite the unquestioned hardiness of the parents, hybrids themselves are sometimes damaged by a hard winter, so in cold areas take the precaution of tying the new canes together in a bundle before the winter and only release them for tying in properly when the worst weather is past. The bundled canes protect each other. Like blackberries, prune the hybrid berries by cutting out the fruited canes soon after fruiting is over. Hybrid cane fruits include the tasty loganberry, the tayberry, sunberry, marionberry, boysen-berry and the recently introduced tummelberry.

Support systems for cane fruits

Summer raspberries, hybrids and blackberries will all grow against a wall or fence or trained to a post and wire system in the open. Raspberries are normally grown in the open tied to two horizontal wires 2ft(60cm) and 4-5ft(1.2-1.5m) high. More vigorous cane fruits need wires 3, 4, 5 and 6ft (1, 1.2, 1.5 and 2m) high. Tie the canes to the wires using the most appropriate system.

Below: The training method you choose will depend on the kind of fruit you are growing. They vary greatly in vigor and the stiffness of their canes. This thornless boysenberry has tall, supple canes so the two-way rope system is the best way to train it.

Although ripe enough for making jellies, these fruits are not yet fit for dessert; they should be darker.

Right: Once established, the tayberry is a very heavy cropper. Crops are up to twice the weight of the loganberry and the plants are hardier. The straw mulch suppresses weeds and helps the soil to retain adequate moisture.

Left: The tayberry is one of the comparatively new race of hybrids. They are similar in flavor, but vary in hardiness, vigor and appearance.

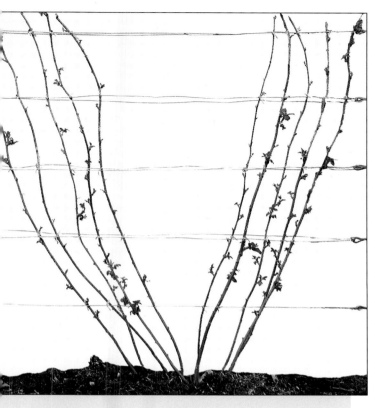

Above: *For less vigorous hybrids, such as this loganberry, a fan system of training is ideal. The new canes are trained up the gap in the center between next year's fruiting canes.*

Above: *The 'Oregon Thornless' blackberry is one of the tastiest thornless varieties and a most attractive plant, with its deeply cut leaves. It is quite vigorous.*

Above: *'Fantasia' is immensely vigorous and thorny, but crops prodigiously and has an excellent flavor. It freezes well, but some berries are apt to go red on freezing.*

Right: *Some of the most vigorous blackberries produce only one or two new canes each year. To fill the allotted space, nip out the top of the new canes when they are 5ft(1.5m) tall to induce side shoots to form. This increases the length of fruiting cane and, therefore, also the crop.*

Grape vines

The only problem with growing grapes in temperate climates is that good-quality dessert varieties are unreliable outdoors and need the protection of a greenhouse. The popular greenhouse variety 'Black Hamburgh' will do quite well outside in warmer areas, as will 'Buckland Sweetwater' (white), but they need a good summer. 'Royal Muscadine' (white) is only slightly lower in quality and more successful. One of the most reliable outdoor varieties is the 'Strawberry Grape' from America. Wine varieties, on the other hand, are no trouble at all. The difficult part of recommending grape varieties is that no two people ever recommend the same ones. However, 'Seyval Blanc' ('Seyve-Villard 5-276') makes a splendid wine and has good disease resistance. 'Reichensteiner' is great; a German grape giving a fresh and fruity wine. Both are fine for gardens.

There is no such thing as a 'best' training method, but outdoors the Guyot system is easy and efficient, especially for wine grapes. After winter planting, cut down the vine to no more than three buds and allow just one shoot to develop. In winter cut this back to four buds. Subsequently, allow just the top two buds to grow into shoots. Next winter, shorten one shoot to two buds, the other to six buds. Bend over the latter shoot and tie it to a supporting wire. It is from this that the fruiting shoots will grow.

The main problem with all grapes is the fungus disease mildew, which can be very damaging and may ruin the crop. It first appears as a white, powdery covering on the newly opened leaves in the spring. It spreads to other leaves as they open and later to the fruitlets. Once the disease reaches those, all is lost. Provide as much ventilation to greenhouse grapes as the temperature will allow and spray all grapes with an appropriate fungicide from early spring onwards.

1 A single cordon grape vine after winter pruning. All the shoots that grew in the previous growing season have been cut back to just one bud.

2 The vine is growing well and is flowering. It is ready for its first summer trim, which prevents excessive growth and favors the fruitlets.

3 *Nip back the tops, leaving two to four leaves beyond the flower truss. This reduces the likelihood of mildew, as well as improving the fruit.*

4 *This shoot has been stopped at four leaves beyond the flower truss and a minor truss has developed. Remove it, along with any other side shoots that grow from a fruiting shoot.*

Above: 'Seyval' ('Seyval Blanc' or 'Seyve-Villard 5-276') is a good and reliable white wine grape. It bears regular and heavy crops, making a delicate wine.

Guyot training system

The Guyot training system is used more for wine grapes than dessert. The example below shows a vine after winter pruning. This is the Double Guyot system, where two fruiting canes are retained.

Above: This vertical cordon of black wine grapes is doing well on a warm and sunny wall. The crop is ripening and all looks set for a good harvesting.

Left: 'Black Hamburgh' is the most reliable black dessert grape for gardeners. It is first-rate in a greenhouse and does well outside, too.

Growing figs outdoors

Figs are Mediterranean fruits that will often grow very well in cool temperate climates, even though they are on the borderline of hardiness. As such, they do best when roughly fan-trained against a warm and sunny wall or fence. They have no particular soil preference, but do not plant them in very rich soil, otherwise they will put on excessive growth at the expense of the fruit. Figs have a strange annual life in that tiny fruitlets develop during late summer, overwinter and mature in the following summer. However, only those fruitlets that are formed after late summer will survive the winter and grow on to maturity; the larger ones die off. You should, therefore, pinch out all fruitlets that appear before late summer. This will result in just two or three tiny figlets at the end of each new shoot by the onset of winter. To encourage this, 'stop' all new shoots at five to six leaves until midsummer.

There are a great many more varieties of fig than most people imagine, but 'Brown Turkey' is the most reliable one for outdoors. It is also an excellent variety in other respects, being the hardiest of all - though it still performs best grown against a sunny wall - and is a heavy cropper with a delicious flavor. The fruits are browny purple with a red interior. 'Brunswick' is another good outdoor variety. It is yellowish green with a tinge of brown on the side that catches the sun. It, too, is red inside. Figs may also be grown very successfully in containers, where they make interesting and attractive patio plants. Start them off in a pot that will just accommodate the root system and only repot them when it becomes difficult to keep them watered sufficiently in the summer. They can be grown either as dwarf bushes or roughly fan-trained to a cane frame.

Above: The developing figs as they look in summer. All the smaller ones should have been removed.

Below: Most varieties ripen during late summer and early fall outside. Leave them on the tree until they are absolutely ripe, even splitting. These are fruits of Ficus carica.

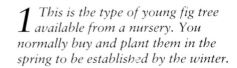

1 This is the type of young fig tree available from a nursery. You normally buy and plant them in the spring to be established by the winter.

Retain just the two
lowest fruits. If allowed
to ripen, the rest will
weaken the tree.

Figs under cover

Figs are ideally suited to being
grown in a greenhouse, mainly
because it considerably extends
the choice of varieties. There is
nothing wrong with the
standard outdoor varieties, but
others are better. 'Rouge de
Bordeaux' is possibly top of the
list, but few nurseries stock it.
'Bourjasotte Grise' and 'Negro
Largo' are virtually
indistinguishable from it and
easier to obtain. Depending on
the space available, grow the
trees fan-trained, either against
a cane frame or on the wall of a
lean-to greenhouse. In either
event, they perform best if
planted in the ground and not
grown in a container.

Below: Fan-trained fig trees against a
sunny wall in early summer. Fruitlets
that overwintered will be swelling.
Pinch out those that are just forming.

2 In spring, reduce the height of the
young tree to about 12in(30cm) to
encourage side shoots to develop. The
fruitlets have still to be removed.

3 This is going to be a fan-trained
tree so make a cane frame and
push it into the ground. As they grow,
tie the new shoots to the frame.

Index to Plants

Page numbers in **bold** indicate major text references. Page numbers in *italics* indicate captions and annotations to photographs. Other text entries are shown in normal type.

Credits

The majority of the photographs featured in this book have been taken by Neil Sutherland and are © Colour Library Books. The publishers wish to thank the following photographers for providing additional photographs, credited here by page number and position on the page, i.e. (B)Bottom, (T)Top, (C)Center, (BL)Bottom left, etc.

Peter Blackburne-Maze: 67(inset BR,CR,CB), 68(TR,BR), 69(T,BL,BR), 72(TR), 73(BL,BR), 74(TL,CL), 94(BR), 95(BL), 99(TC,TR), 100(T), 101(TR,BL,BR)

Eric Crichton: 10, 15(CR,BR), 19(BR), 26(BL), 28(BL,BR), 37(BR), 47(TL), 49(L,TR), 50(TR), 73(TL,TR), 82(BR), 85(BR), 95(CL), 100(B), 101(TC), 103(TR), 105(BR)

John Glover: Half-title, 8, 15(TR), 35(TL), 39(BR), 68(BL), 82(BC), 83(TR), 104(T)

Photos Horticultural Picture Library: 38(TR), 39(TR), 51(TL), 83(BL), 99(BL), 104(BL)

Acknowledgments

The publishers would like to thank the following people and organizations for their help during the preparation of this book: The Royal Horticultural Society Garden at Wisley, Surrey; Brinsbury College - The West Sussex College of Agriculture and Horticulture; John Nash and Sally Cave at Costrong Farm, West Sussex; Lynn Hutton at Secrets Farm Shop, Surrey; The National Institute of Agricultural Botany, Cambridge; Murrells Nursery, West Sussex; Jill Coote.